SPEECH SEPARATION BY HUMANS AND MACHINES

SPEECH SEPARATION BY HUMANS AND MACHINES

Edited by

Pierre Divenyi
East Bay Institute for Research and Education

KLUWER ACADEMIC PUBLISHERS

Distributors for North, Central and South America:
Kluwer Academic Publishers
101 Philip Drive
Assinippi Park
Norwell, Massachusetts 02061 USA
Telephone (781) 871-6600
Fax (781) 681-9045
E-Mail: kluwer@wkap.com

Distributors for all other countries:
Kluwer Academic Publishers Group
Post Office Box 322
3300 AH Dordrecht, THE NETHERLANDS
Telephone 31 786 576 000
Fax 31 786 576 474
E-Mail: services@wkap.nl

 Electronic Services <http://www.wkap.nl>

Library of Congress Cataloging-in-Publication Data

Speech Separation by Humans and Machines
Pierre Divenyi (Ed.)
ISBN 1-4020-8001-8
eISBN 0-387-22794-6

This material is based upon work supported by the National Science Foundation under Grant No. 0345301. Any opinions, findings, and conclusions or recommendations expressed in this material are those of the author(s) and do not necessarily reflect the views of the National Science Foundation.

To Mary, from all of us.

Contents

Contributors

Chris Darwin
Department of Psychology, School of Life Sciences, University of Sussex, United Kingdom.

Elyse S. Sussman
Department of Neuroscience, Albert Einstein College of Medicine, Bronx, New York, NY.

Claude Alain
Rotman Research Institute, Baycrest Centre for Geriatric Care and Department of Psychology, University of Toronto, Ontario.

Peter Cariani
Eaton Peabody Laboratory of Auditory Physiology and Massachusetts Eye & Ear Infirmary, Boston, MA.

Te-Won Lee
Institute for Neural Computation, University of California, San Diego, CA.

Bhiksha Raj
Mitsubishi Electric Research Laboratories, Cambridge, MA.

Michael Seltzer
Microsoft Research, Redmond, WA.

Manuel Jesus Reyes-Gomez
Columbia University, New York, NY.

Paris Smaragdis
Mitsubishi Electric Research Laboratories, Cambridge, MA.

Sam T. Roweis
Department of Computer Science, University of Toronto, Ontario.

Richard M. Stern
Carnegie Mellon University, Pittsburgh, PA.

Toshio Irino
Faculty of Systems Engineering, Wakayama University, Wakayama, Japan.

Roy D. Patterson
CNBH, Physiology Department, Cambridge University, Cambridge, United Kingdom.

Hideki Kawakhara
Faculty of Systems Engineering, Wakayama University, Wakayama, Japan.

DeLiang Wang
Department of Computer Science and Engineering and Center of Cognitive Science, The Ohio State University, Columbus, OH.

Malcolm Slaney
IBM Almaden Research Center, San Jose, CA.

Guy J. Brown
Department of Computer Science, University of Sheffield, United Kingdom.

Kalle J. Palomäki
Laboratory of Acoustics and Audio Signal Processing, Helsinki University of Technology, Finland.

Nat Durlach
Hearing Research Center, Boston University, Boston, MA and Sensory Communication Group, Research Laboratory of Electronics, Massachusetts Institute of Technology, Cambridge, MA.

Alain de Cheveigné
Ircam-CNRS, Paris, France.

Douglas S. Brungart
Air Force Research Laboratory, Arlington, VA.

Pierre L. Divenyi
Veterans Affairs Medical Center and East Bay Institute for Research and Education, Martinez, CA.

Ziyou Xiong
Dept. of Computer And Electrical Engineering, University Of Illinois At Urbana-Champagne, IL.

Thomas S. Huang
Dept. of Computer And Electrical Engineering, University Of Illinois At Urbana-Champagne, IL.

Daniel P.W. Ellis
LabROSA, Columbia University, New York NY.

Martin Cooke
Department of Computer Science, University of Sheffield, United Kingdom.

Foreword

There is a serious problem in the recognition of sounds. It derives from the fact that they do not usually occur in isolation but in an environment in which a number of sound sources (voices, traffic, footsteps, music on the radio, and so on) are active at the same time. When these sounds arrive at the ear of the listener, the complex pressure waves coming from the separate sources add together to produce a single, more complex pressure wave that is the sum of the individual waves. The problem is how to form separate mental descriptions of the component sounds, despite the fact that the "mixture wave" does not directly reveal the waves that have been summed to form it.

The name *auditory scene analysis* (ASA) refers to the process whereby the auditory systems of humans and other animals are able to solve this mixture problem. The process is believed to be quite general, not specific to speech sounds or any other type of sounds, and to exist in many species other than humans. It seems to involve assigning spectral energy to distinct "auditory objects" and "streams" that serve as the mental representations of distinct sound sources in the environment and the patterns that they make as they change over time. How this energy is assigned will affect the perceived number of auditory sources, their perceived timbres, loudnesses, positions in space, and pitches. Indeed, every perceived property studied by psychoacoustics researchers seems to be affected by the partitioning of spectral energy. While the name ASA refers to the competence of humans and other animals, the name *computational* auditory scene analysis (CASA) refers to the attempt by scientists to program computers to solve the mixture problem.

In 2003, Pierre Divenyi put together an interdisciplinary workshop that was held in Montreal that autumn, a meeting focused on the topic of how to separate a speech signal from interfering sounds (including other speech). It is obvious why this topic is so important. Right now speech recognition by computers is a delicate process, easily derailed by the presence of interfering sounds. If methods could be evolved to focus recognition on just those components of the signal that came from a targeted source, recognition would be more robust and usable for human-computer interaction in a wide variety of environments. Yet, albeit of overwhelming importance, speech separation represents only a part of the more general ASA problem, the study of which may shed light on issues especially relevant to speech understanding in interference. It was therefore appropriate that Divenyi assembled members of a number of disciplines working on the problem of the separation of concurrent sounds: experimental psychologists studying how ASA was done by people, both for speech and non-speech sounds, neuroscientists interested in how the brain deals with sounds, as well as computer scientists and engineers developing computer systems to solve the problem. This book is a fascinating collection of their views and ideas on the problem of speech separation.

My personal interest in these chapters is that they bring to forefront an argument of special import to me as a cognitive psychologist. This argument, made by CASA researchers, is that since people can do sound separation quite well, a better understanding of *how* they do it will lead to better strategies for designing computer programs that can solve the same problem. Others however, disagree with this argument, and want to accomplish sound segregation using any powerful signal-processing method that can be designed from scientific and mathematical principles, without regard for how humans do it. This difference in strategy leads one to ask the following question: Will one approach ultimately wipe out the other or will there always be a place for both? Maybe we can take a lesson from the ways in which humans and present-day computer systems are employed in the solving of problems. Humans are capable of solving an enormous variety of problems (including how to program computers to solve problems). However, they are slow, don't always solve the problems, and are prone to error. In contrast, a computer program is typically designed to carry out a restricted range of computations in a closed domain (e.g., statistical tests), but can do them in an error-free manner at blinding speeds. It is the "closedness" of the domain that permits a strict algorithmic solution, leading to the blinding speed and the absence of error. So we tend to use people when the problems reside in an "open" domain and computers when the problem domain is closed and well-defined. (It is possible that when computers become as all purpose and flexible in their thought as humans, they will be as slow and as subject to error as people are.)

The application of this lesson about general-purpose versus specialized computation to auditory scene analysis by computer leads to the conclusion that we should use general methods, resembling those of humans, when the situation is unrestricted – for example when both a robotic listener and a number of sound sources can move around, when the sound may be coming around a corner, when the component sounds may not be periodic, when substantial amounts of echo and reverberation exist, when objects can pass in front of the listener casting acoustic shadows, and so on. On the other hand, we may be able to use faster, more error-free algorithms when the acoustic situation is more restricted.

If we accept that specialized, algorithmic methods won't always be able to solve the mixture problem, we may want to base our general CASA methods on how people segregate sounds. If so, we need a better understanding of how human (and animal) nervous systems solve the problem of mixture.

Achieving this understanding is the role that the experimental psychologists and the neuroscientists play in the CASA enterprise.

The present book represents the best overview of current work in the fields of ASA and CASA and should inspire researchers with an interest in sound to get involved in this exciting interdisciplinary area. Pierre Divenyi deserves our warmest thanks for his unstinting efforts in bringing together scientists of different orientations and for assembling their contributions to create the volume that you are now reading.

Albert S. Bregman
Professor Emeritus of Psychology
McGill University

Preface

Speech against the background of multiple speech sources, such as crowd noise or even a single talker in a reverberant environment, has been recognized as the acoustic setting perhaps most detrimental to verbal communication. Auditory data collected over the last 25 years have succeeded in better defining the processes necessary for a human listener to perform this difficult task. The same data has also motivated the development of models that have been able to increasingly better predict and explain human performance in a "cocktail-party" setting. As the data showed the limits of performance under these difficult listening conditions, it became also clear that significant improvement of speech understanding in speech noise is likely to be brought about only by some yet-to-be-developed device that automatically separates the speech mixture, enhances the target source, and filters out the unwanted sources. The last decade has allowed us to witness an unprecedented rush toward the development of different computational schemes aimed at achieving this goal.

It is not coincidental that computational modelers started looking at the problem of auditory scene analysis largely in response to Albert Bregman's book (1990) which appeared to present a conceptual framework suitable for computational description. Computer scientists, working at different laboratories, rose to the challenge and designed algorithms that implemented human performance-driven schemes, in order to achieve computational separation of simultaneous auditory signals – principally with speech as the target. Computational Auditory Scene Analysis (CASA) has gained adepts and enjoyed exposure ever since the first CASA workshop in 1995. Nevertheless, it became clear early on that literal implementations of the systems described by Bregman could not achieve separation of speech signals without blending them with methods taken from other frameworks. Driven by different objectives, such as analysis of EEG responses and neuroimaging data in addition to analysis of acoustic signals, Independent Component Analysis (ICA) appeared at about the same time, motivated by Bell's and Sejnowski's seminal article (1995). Early implementations of ICA as the engine for blind separation of noisy speech signals gave impressive results on artificial mixtures and served as a starting point for numerous initiatives in applying it to the separation of multichannel signals. However, limitations of ICA also became apparent, mainly when applied to real-world multichannel or, especially, to single-channel mixtures. At the same time, advances in speech recognition through the 1990s have led researchers to consider challenging scenarios, including speech against background interference, for which signal separation seems a promising approach, if not a much-needed tool. In a reciprocal fashion, the influence of speech recognition has brought to CASA statistical pattern recognition and machine learning ideas that exploit top-down constraints, although at the cost of using very large training sets to tune parameters. Top-down techniques have been applied in conjunction with ICA-based separation methods as well, contemporaneously with neurophysiological findings that have been uncovering an ever larger role of corticofugal efferent systems for speech understanding by human listeners.

A cursory survey of computational separation of speech from other acoustic signals, mainly other speech, strongly suggests that the current state of the whole field is in a flux: there are a number of initiatives, each based on an even larger number of theories, models, and assumptions. To a casual observer it seems that, despite commendable efforts and achievements by many researchers, it is not clear where the field is going. At the same time, despite an accelerating increase in investigations by neuroscientists that have led to characterizing and mapping more and more of the auditory and cortical processes responsible for speech separation by man, our understanding of the entire problem still seems to be far away. One possible reason for this generally unsatisfactory state of affairs could well be that investigators working in separate areas still seldom interact and thus cannot learn from each other's achievements and mistakes.

In order to foster such an interaction, an invitational workshop was held in Montreal, Canada, over the weekend of October 31 to November 2, 2003. The idea of the workshop was first suggested by Dr. Mary P. Harper, Director of the Human Language and Communication Program at the National Science Foundation, who stood behind the organizers and actively helped their efforts at every step. Her enthusiastic interest in the topic was also instrumental in obtaining, within an atypically short time period, sponsorship by the Foundation's Division of Intelligent Information Systems at the Directorate for Computer and Information Science and Engineering,. The workshop was attended by some twenty active presenters — a representative sample of experts of computational, behavioral, and neurophysiological speech separation working on different facets if the general problem area and using different techniques. In addition, representatives of a number of funding agencies also attended. Interspersed with presentations of the experts' work, there were periods of planned discussion that stimulated an intensive exchange of ideas and points of view. It was the unanimous opinion of all those present that this exchange opened new alleys toward a better understanding of each other's research as well as where the whole field stood. The discussions also identified directions most beneficial and productive for future work on speech separation.

The workshop was, to our knowledge, the first of its kind and also timely. Indeed, prior to the workshop, an overlook of the various methods would have made the observer conclude that the sheer volume of contemporary work on speech separation had been staggering, that the work had its inherent limitations regardless of the philosophy or technological approach that it followed, and that at least some of these limitations might be overcome by adopting a common ground. This latter proposition, however, had the prerequisite that proponents of different approaches get together, present their methods, theories, and data, discuss these openly, and attempt to find ways to combine schemes, in order to achieve the ultimate goal of separating speech signals by computational means. Reaching this objective was the Montreal workshop's main purpose. Another goal was to encourage investigators from different fields and adepts of different methods to explore possibilities of attacking the problem jointly, by forming collaborative ventures between computer scientists working on the problem from different directions, as well as engaging in interdisciplinary collaboration between behavioral, neurobiological, and computer scientists dedicated to the study of speech. Finally, thanks to the presence of program administrators from different Federal agencies at the workshop, we hoped to create some momentum toward the development of intra- and interagency programs aimed at better understanding speech separation from the points of view of computer science, behavioral science, and neuroscience.

This book is a collection of papers written by the workshop's participants. Although the chapters follow the general outline of the talks that their authors gave in Montreal, they have generally outgrown the material presented at the workshop, both in their scope and their orientation. Thus, the present book is not merely another volume of conference proceedings or a set of journal articles gone astray. Rather, it is a matrix of data, background, and ideas that cuts across fields and approaches, with the unifying theme of separation and interpretation of speech corrupted by a complex acoustic environment. Over and above presenting facts and accomplishments in this field, the landscape that the book wishes to paint also highlights what is still missing, unknown, and un-invented. It is our hope that the book will inspire the reader to learn more about speech separation and even will be tempted to join the rows of investigators busy trying to fill in the blanks.

The chapters of the book cover three general areas: neurophysiology, psychoacoustics, and computer science. However, several chapters are difficult to categorize because they straddle across two fields. For this reason, although the book is not divided into clearly delimited sections, the reader should recognize a thread that ties the chapters together and attempts to tell a coherent story. Darwin's send-off chapter introduces speech separation from the behavioral point of view, showing human capabilities and limits. These capabilities are examined from the viewpoint of neurophysiology in the chapters by Sussman and Alain, both showing how auditory scene analysis can be studied by observing evoked cortical potentials in alert human subjects. Cariani's chapter reviews data on recordings from the cat's peripheral auditory system and analyses the data to show how, even at a level as low as the auditory nerve, the system is able to extract pitch information and perform a complex computation—decorrelation—that may account for the separation of the voice of two simultaneous talkers. Decorrelation also underlies ICA, a procedure that can be successfully used for blind separation of simultaneous speech streams, as Lee's chapter demonstrates—the adjective "blind" referring to a statistical method that takes no assumption with regard to the origin or nature of the source. A different approach is adopted in the chapter by Raj, Seltzer, and Reyes, who propose a speech separation system that uses beamforming to enhance the output of a multiple-microphone array but that also takes advantage of previously learned statistical knowledge about language. An inferential approach is taken by Roweis in his chapter, based on probabilistic estimation of the input source, which is shown to help denoising, separating, and estimating various parameters of speech. Stern is taking yet another different route: applying auditory models to recover speech, both by separation and by recognition—a hybrid bottom-up and top-down approach. Irino, Patterson, and Kawahara show how speech separation can be achieved by using a combination of two sophisticated signal processing methods: the auditory image method (AIM) and STRAIGHT, a pitch-synchronous formant tracking method. This latter method is further described by Kawahara and Irino showing how STRAIGHT, a method consistent with auditory processing, can also lead to the separation of speech signals even by itself. Auditory models are the explicitly stated base of CASA, the focus of several successive chapters. The one by Wang and Hu suggests that separation of speech from a noisy background can be achieved by applying an estimated optimal binary mask, but they also point out that no unambiguous definition of CASA currently exists. Slaney, in basic agreement with this proposition, presents a historical overview of CASA and discusses its advantages as well as its flaws. One undisputable positive trait of CASA is its ability to spatially segregate signals by adopting features of binaural hearing. Brown and Palomäki show that these features can be applied to recover speech presented in a reverberant environment—a notoriously difficult situation. Durlach, a proponent of the now-classic equalization-and-cancellation (EC) model of binaural hearing, shows how sound localization by listeners can be used for not only separating but also interpreting mixed multiple speech signals. De Cheveigné takes the EC concept further to demonstrate that the basic concept of this model can also be applied to separate the mixed speech of several talkers on the basis of fundamental frequency pitch. These multiple speech signals can be regarded as examples of one stream masking the information, rather than the energy, in the other stream or streams, as Brungart's chapter shows. Divenyi takes an analytic approach to informational masking to show how a given feature in a speech stream—envelope fluctuations or formant trajectories—can be masked by random sequences of the same feature in another stream. The chapter by Xiong and Huang steps out of the acoustic domain and shows how visual and auditory information by talkers speaking simultaneously interact and how this interaction can be successfully used by humans and machines to separate multiple speech sources. Ellis addresses one of the ubiquitous problems of speech recovery by machines: performance evaluation. The final chapter by Cooke takes a broad view of speech separation. Looking through the lens of someone committed to uncovering the mysteries of ASA and CASA, he proposes that time-frequency information in badly degraded portions of speech may be recovered by glimpsing at those regions where this information is not as severely degraded.

While this round-up does no justice to the twenty-one chapters of the book, it hopes to convey two important points. First, we want to emphasize that at present the field of speech separation by human and machines is complex and replete with ill-defined and un-agreed-upon concepts. Second, despite this, we want to express our hope that significant accomplishments in the field of speech separation may be come to light in the not-so-distant future, propelled by a dialog between proponents of different approaches, The book serves as an illustration of the complementarity, on the one hand, and the mutual overlap, on the other, between these approaches and its sheer existence suggests that such a dialog is possible.

The volume would not exist without the contribution of many. Dan Ellis and DeLiang Wang helped me organize the 2003 Montreal workshop. The workshop was supported by the National Science Foundation and by a contribution from Mitsubishi Electric Research Laboratories. I also want to acknowledge the efficient technical support of the workshop by Reyhan Sofraci and her staff at the Royal Crown Plaza Hotel in Montreal. Theresa

Azevedo, President of the East Bay Institute for Research and Education threw all her energy behind the actual realization of the workshop and production of this book. Joanne Hanrahan spent countless hours assembling, editing, and formatting the chapters. I also want to thank Connie Field, my wife, for tolerating my preoccupied mind before and during the preparation of this book. Lastly and most of all, however, I want to express my gratitude to Mary Harper, whose enthusiastic support of multidisciplinary research on speech separation was responsible for the workshop and has created the impetus for the book.

Pierre Divenyi

Bell, A.J. and Sejnowski, T.J., 1995, An information maximisation approach to blind separation and blind deconvolution. *Neural Computation,* 7(**6**), 1129-1159.

Bregman, A.S., 1990, *Auditory scene analysis: The perceptual organization of sound.* Cambridge, Mass.: Bradford Books (MIT Press).

Chapter 1

Speech Segregation: Problems and Perspectives.

Chris Darwin
Department of Psychology
School of Life Sciences, University of Sussex
cjd@biols.susx.ac.uk

1   INTRODUCTION

Although it has become increasingly obvious in the speech recognition literature that the recognition of speech mixed with other sounds poses difficult problems, it is not clear what the appropriate approach to the problem should be. On the one hand, recognition could be based on matching an internally generated mixture of sounds to the input (mixed) signal (Varga and Moore, 1990). Such an approach requires good generative models not only for the target sounds but also for interfering sounds. On the other hand, the (potentially multiple) sound sources could be segregated prior to recognition using low-level grouping principles (Bregman, 1990). Such an approach could in principle be a more general solution to the problem, but depends on how well the actual sound sources can be separated by grouping principles established generally with simple musical sounds. Is speech a single sound source in these terms? What is a sound source anyway?

Like visual objects, sound sources can be viewed as being hierarchically organised (the Attack of the Lowest note of the Chord of the Leader of the 1° Violin section of the Orchestra), indeed much of the point of music is the interplay between homophonic and polyphonic perceptions. Speech can be regarded as a mixture of simpler sound sources: Vibrating vocal folds, Aspiration, Frication, Burst explosion, Ingressive Clicks. For our native tongue, these multiple sources integrate into "speech", but for talkers of non-click languages, clicks may sound perceptually separate failing to integrate into the sound–stream of the speech. Even a steady vocalic sound produced by the vocal tract acting as a single tube can perceptually fragment into two sound sources – as when a prominent high harmonic is heard as a separate whistle in Tuvan throat music (Levin and Edgerton, 1999).

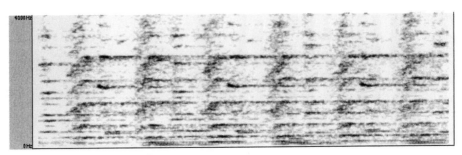

Figure 1.1. Original (Berlioz Symphonie Fantastique)

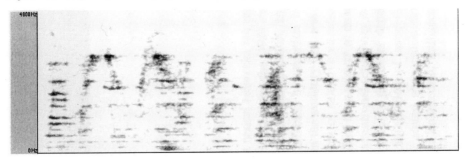

Figure 1.2. LPC filtered (Berlioz + "where were you a year ago?")

One can even question whether the notion of a sound *source* is appropriate for speech. As Melvyn Hunt demonstrated many years ago, we can hear a ghostly voice emerging through the orchestra when the orchestral sound replaces laryngeal excitation in an LPC resynthesis of a sentence. The example above is of the ubiquitous "Where were you a year ago" spoken twice.

Here it is perhaps the joint transfer function of multiple sound sources which is giving the speech percept. Attempts at source separation could simply make matters worse!

In practice of course the speech we wish to listen to is not the result of filtering an orchestra, and is often spatially well–localised. Although we *can* attend to one voice in a mixture from a single spatial location, spatial separation of sound sources helps. How do listeners use spatial separation? A major factor is simply that head shadow increases S/N at the ear nearer the target (Bronkhorst and Plomp, 1988). But do localisation cues also contribute to selectively grouping different sound sources?

An interesting difference appears between simultaneous and successive grouping in this respect. For successive grouping, where two different sound sources are interleaved, spatial cues, both individually and in combination provide good segregation – as in the African xylophone music demonstration on the Bregman and Ahad CD (1995).

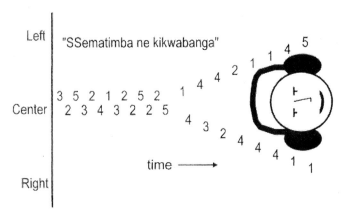

Figure 1.3. Streaming by spatial location in African xylophone music reproduced with permission from Bregman and Ahad (1996).

However, a recent counter-intuitive result is that most inexperienced listeners are surprisingly poor at using the most powerful cue for localising natural speech (interaural–time differences – ITDs) to perform simultaneous grouping of sounds which have no other grouping cues (Culling and Summerfield, 1995, Hukin and Darwin, 1995, Drennan *et al.*, 2003). The significance of this result may be that binaural information about location is pooled across the constituent frequencies of an auditory object, in order to improve the reliability and stability of its perceived position (we don't after all hear different spectral parts of an auditory object in different locations, even under the most difficult listening conditions). This observation, and others, argue for some perceptual grouping preceding the use of binaural cues in estimating the location of an object (Woods and Colburn, 1992, Darwin and Hukin, 1999). The implication for speech segregation by machine, is that spatial cues may be of only limited use (particularly in reverberant conditions) in the absence of other cues that help to establish the nature of the auditory objects.

Different aspects of auditory scene analysis may use spatial cues at different levels. For example, successive separation of interleaved rhythms depends on subjective spatial position, rather than a difference in any one spatial cue (Sach and Bailey, in press), while on the other hand the perceived continuity of a tone alternating with noise depends on the ITD relations between the tone and the noise, not on their respective spatial positions (Darwin *et al.*, 2002).

References

Bregman, A.S., 1990, *Auditory Scene Analysis: the perceptual organisation of sound,* Bradford Books (MIT Press), Cambridge, Mass.

Bregman, A.S. and Ahad, P., 1995, *Compact Disc: Demonstrations of auditory scene analysis* Department of Psychology, McGill University, Montreal.

Bronkhorst, A.W. and Plomp, R., 1988, The effect of head–induced interaural time and level differences on speech intelligibility in noise, *J. Acoust. Soc. Am.* **83**, 1508–1516.

Culling, J.F. and Summerfield, Q., 1995, Perceptual separation of concurrent speech sounds: absence of across–frequency grouping by common interaural delay, *J. Acoust. Soc. Am.* **98**, 785–797.

Darwin, C.J., Akeroyd, M.A. and Hukin, R.W., 2002, Binaural factors in auditory continuity, *Proceedings of the 2002 International Conference on Auditory Display*, July 2–5, 2002, Kyoto, Japan,. pp. 259–262.

Darwin, C.J. and Hukin, R.W., 1999, Auditory objects of attention: the role of interaural time–differences, *J. Exp. Psychol.: Hum. Perc. & Perf.* **25**, 617–629.

Drennan, W.R., Gatehouse, S. and Lever, C., 2003, Perceptual segregation of competing speech sounds: the role of spatial location, *J. Acoust. Soc. Am.* **114**, 2178–89.

Hukin, R.W. and Darwin, C.J., 1995, Effects of contralateral presentation and of interaural time differences in segregating a harmonic from a vowel, *J. Acoust. Soc. Am.* **98**, 1380–1387.

Levin, T.C. and Edgerton, M.E., 1999, The throat singers of Tuva, *Scientific American* September 1999.

Sach, A.J. and Bailey, P.J. (in press) Some aspects of auditory spatial attention revealed using rhythmic masking release, *J. Exp. Psychol.: Hum. Perc. & Perf.*

Varga, A.P. and Moore, R.K., 1990, Hidden Markov Model decomposition of speech and noise, *IEEE International Conference on Acoustics, Speech and Signal Processing*, Albuquerque. pp. 845–848.

Woods, W.A. and Colburn, S., 1992, Test of a model of auditory object formation using intensity and interaural time difference discriminations, *J. Acoust. Soc. Am.* **91**, 2894–2902.

Chapter 2

Auditory Scene Analysis:
Examining the Role of Nonlinguistic Auditory Processing In Speech Perception

Elyse S. Sussman
Department of Neuroscience, Albert Einstein College of Medicine
esussman@aecom.yu.edu

1   INTRODUCTION

From infancy we experience a complex auditory environment with acoustic information originating from several simultaneously active sources that often overlap in many acoustic parameters. Despite this confluence of sound we are able to hear distinct auditory objects and experience a coherent environment consisting of identifiable auditory events. Analysis of the auditory scene (Auditory Scene Analysis or ASA) involves the ability to integrate those sound inputs that belong together and segregate those that originate from different sound sources (Bregman, 1990). Accordingly, integration and segregation processes are two fundamental aspects of ASA. This chapter focuses on the interaction between these two important auditory processes in ASA when the sounds occur outside the focus of one's attention.

A fundamental aspect of ASA is the ability to associate sequential sound elements that belong together (integration processes), allowing us to recognize a series of footsteps or to understand spoken speech. Auditory sensory memory plays a critical role in the ability to integrate sequential information. Think about how we understand spoken speech. Once each word is spoken, only the neural trace of the physical sound information remains. Auditory memory allows us to access the series of words that were spoken, and connect them to make meaning of the individual words as a unit. Transient auditory memory has been estimated to store information for a period of time at least 30 s (Cowan, 2001). In understanding how this memory operates in facilitating ASA, it is important to also understand the relationship between the segregation and integration processes. The question of how sequential sound elements are represented and stored in auditory memory can be explored using event–related brain potentials (ERPs).

2   EVENT–RELATED BRAIN POTENTIALS

ERPs provide a non-invasive measure of cortical brain activity in response to sensory events. We can gain information about the timing of certain cognitive processes evoked by a given sound because of the high temporal resolution (in the order of milliseconds) of the responses that are time-locked to stimulus events. ERPs provide distinctive signatures for sound change detection. Of particular importance is the mismatch negativity (MMN) component of ERPs, which reflects sound change detection that can be elicited even when the sounds have no relevance to ongoing behavior (Näätänen *et al.*, 2001). MMN is generated within auditory cortices (Giard *et al.*, 1990, Javitt *et al.*, 1994) and is usually evoked within 200 ms of sound change, representing an early process of change detection. MMN generation is dependent upon auditory memory. Evidence that activation of NMDA receptors plays a role in the MMN process supports the notion that the underlying mechanisms of this cortical auditory information processing network involve sensory memory (Javitt *et al.*, 1996). The neural representations of the acoustic regularities (often called the "standard"), which are extracted from the ongoing sound sequence, are maintained in memory and form the basis for the change detection process. Incoming sounds that deviate from the neural trace of the standard elicit MMN. Thus, the presence of the MMN can also be used to ascertain what representation of the standard was stored in memory.

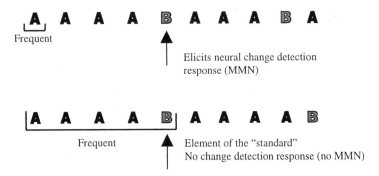

Figure 2.1 In the top panel, a typical auditory oddball paradigm is shown. An "oddball" or infrequently–occurring sound (represented by the letter B) is randomly presented amongst a frequently repeating sound (represented by the letter A). The oddball (B) elicits MMN in this context. In the bottom panel, the same ratio of B to A sounds (1:5) is presented, but instead of presenting B randomly, it is presented every fifth tone in the sequence. *If* the brain detects the frequently–repeating 5-tone pattern then MMN is *not* elicited by the "oddball" (B) tone in this context. See text for further details.

3   MISMATCH NEGATIVITY IS CONTEXT–DEPENDENT

MMN can be used to probe neural representations of the regularities extracted from the ongoing sound input exactly because the response to a particular sound is based upon the memory of the previous sounds. This is illustrated in Figure 2.1. Using a simple auditory oddball paradigm, in which an "oddball" (or infrequently occurring sound) is presented randomly among frequently repeating sounds, the *oddball* elicits MMN when it is detected as deviating (e.g., in frequency, intensity, duration, or spatial location) from the frequently repeating sound. In the top panel of Figure 2.1, a typical auditory oddball paradigm is shown. The letter "A" represents a tone of one frequency and the letter "B" represents a tone of a different frequency. The oddball (B) elicits MMN because it has a different frequency than that of the standard (A). In the bottom panel of Figure 2.1, the same ratio of B to A sounds is presented, but instead of presenting "B" randomly, it is presented every fifth tone in the sequence. *If* the brain detects the regularity (the 5-tone repeating pattern A-A-A-A-B-A-A-A-A-B...) then no MMN is elicited by the B tone (the "oddball"). This is because when the 5-tone pattern is detected, the B tone is part of the standard repeating regularity (Sussman *et al.*, 1998a, 2002a); it is not a deviant. Accordingly, it is important to notice that MMN is not simply elicited when there is a frequent and an infrequent tone presented in the same sequence. MMN generation depends on detection and storage of the regularities in the sound stimulation. The detected regularities provide the auditory context from which deviance detection ensues. Thus, MMN is *highly* dependent upon the context of the stimuli, either when the context is detected without attention focused on the sounds (Sussman *et al.*, 1999, 1998a, 2002b, 2003, Sussman and Winkler, 2001) or when the context is influenced by attentional control (Sussman *et al.*, 1998b, 2002a).

4   SEGREGATION PROCESSES AND ERPS

The acoustic information entering one's ears is a mixture of all the sound in the environment, without separation. A key function of the auditory system is to disentangle the mixture and construct, from the simultaneous inputs, neural representations of the sound events that maintain the integrity of the original sources (Bregman, 1990). Thus, decomposing the auditory input is a crucial step in auditory information processing, one that allows us to detect a single voice in a crowd or distinguish a voice coming from the left or right at a cocktail party. The process of disentangling the sound to sources (ASA) plays a critical role in how we experience the auditory environment. There is now considerable ERP evidence to suggest that auditory memory can hold information about multiple sound streams independently and that the segregation of auditory input to distinct sound streams can occur without attention focused on the sounds (Sussman *et al.*, submitted-a, Sussman *et al.*, submitted-b, Sussman *et al.*, 1999, Ritter *et al.*, 2000, Winkler *et al.*, 2003), even though there remains some controversy about whether or not attention is needed to segregate the sound input (Botte *et al.*, 1997, Bregman, 1990, Brochard *et al.*, 1999, Carlyon *et al.*, 2001, Macken *et al.*, 2003, Sussman *et al.*, 1999, Winkler *et al.*, 2003). Functionally, the purpose for automatic grouping processes would be to facilitate the ability to select information. In this view, the role of attention is not to specifically organize the sounds, some organization of the input is calculated and stored in memory without attention. Attentional resources are needed for identifying attended source patterns, which is essential for understanding speech or for appreciating music. Attention, however, can modify the organization of the sound input (Sussman *et al.*, 1998b, Sussman *et al.*, 2002a), which then influences how the information is stored and used by later processes (e.g., the MMN process).

5   INTEGRATION PROCESSES AND ERPS

The perception of a sound event is often determined by the sounds that surround it even when the sounds are not in close temporal proximity. Changes in the larger auditory context have been shown to affect processing of the individual sound elements (Sussman *et al.*, 2002b, Sussman and Winkler, 2001). The ability of the auditory system to detect contextual changes (such as the onset or cessation of sounds within an ongoing sound sequence) thus plays an important role in auditory perception. The dynamics of context-change detection were investigated in Sussman and Winkler (2001) in terms of the contextual effects on auditory event formation when subjects had no task with the sounds. The presence or absence of "single deviants" (single frequency deviants) in a sound sequence that also contained "double deviants" (two successive frequency deviants) created different contexts for the evaluation of the double deviants (see Figure 2.2). The double deviants were processed either as unitary events (one MMN elicited by them) or as two successive events (two MMNs elicited by them) depending on which context they occurred in (Sussman *et al.*, 2002b). In Sussman and Winkler, the context was modified by the onset or cessation of the single deviants occurring within a continuous sequence that also contained the double deviants. The time course of the effects of the contextual changes on the brain's response to the double deviants was assessed by whether they elicited one MMN (in the Blocked context) or two MMNs (in the Mixed context). The change of response to the double deviants from one to two MMNs or from two to one MMN did not occur immediately. It took up to 20 s after the onset or cessation of the single deviants before the MMN response to the double deviants reflected the context change. This suggests that there is a biasing of the auditory system to maintain the current context until enough evidence is accumulated to establish a true change occurred, thus avoiding miscalculations in the model of the ongoing sound environment. The results demonstrated that the auditory system maintains contextual information and monitors for sound changes within the current context, even when the information is not relevant for behavioral goals.

6 RELATIONSHIP BETWEEN SEGREGATION AND INTEGRATION IN ASA

Two important conclusions can be ascertained from the ERP results of the segregation and integration processes discussed above. 1) Segmentation of auditory input can occur without attention focused on the sounds. 2) Within–stream contextual factors can influence how auditory events are represented

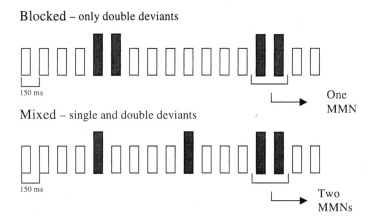

Figure 2.2 The white bars represent a frequently repeating tone (standard) and the black bars represent an infrequently occurring tone with a different frequency than the standard (deviant). In the top panel, a Blocked context is shown – every time a frequency deviant occurs a second deviant follows it (called "double deviants"). The double deviants in this context elicit one MMN. In the bottom panel, a Mixed context is shown – the same ratio of deviants to standards is presented as in the blocked context, except that single deviants are randomly mixed in with the double deviants. The double deviants in this context elicit two MMNs.

in memory. How do these two processes – segregation of high from lowtones, and integration of double deviants within a stream – interact when called upon to function together? This was recently tested in a study investigating whether contextual influences on auditory event formation would occur when two concurrent sound streams were present in the auditory input (Sussman, submitted). It was hypothesized that the acoustic characteristics of the input (the stimulus–driven cues) would be used to separate sounds to distinct sources *prior to* the integration of elements and the formation of sound events. This is what was found. Within–stream contextual influences on event formation were found similarly as were found when one sound stream was presented alone (as in Sussman *et al.*, 2002b). Because the high and low sounds were presented in an alternating fashion, the results indicate that the separation of sounds to streams occurred prior to integration processes. Taken together with previous results (e.g., Sussman *et al.*, 1998b, Sussman *et al.*, 1999, Yabe *et al.*, 2001), there is strong evidence that segregation is an earlier, primitive process than integration that is initially driven by stimulus–characteristics of the acoustic input. This evidence is consistent with animal studies demonstrating that the basic stream segregation mechanisms exist as part of all vertebrates' hearing systems (Fay, 2000, Hulse *et al.*, 1997, Fishman *et al.*, 2001). Integration of sequential elements to perceptual units takes place on the already segregated streams, which would be needed to identify within–stream sound patterns in natural situations that contain acoustic information emanating from multiple sources, making it possible to hear a single speech stream in a crowded room.

The putative timing or sequence of events that has been demonstrated with the ERP results of the studies discussed would essentially operate as follows (the wealth of feedback and parallel processing mechanisms that are also engaged in the neural model are not included here for simplicity). The mixture of sounds enters the auditory system and is initially segregated according to the acoustic characteristics of the input (the frequency, intensity, duration, and spatial location components as well as the timing of the input). Sound regularities are extracted from the input and integration processes, or sound event formation, then operate on the segregated sound streams. The MMN process uses this information (the segregated input and neural representation of the relevant context) as the basis for detecting what has changed in the environment. When attention modifies the initial organization of the sound input, it affects event formation and how the information is represented and stored in memory (Sussman *et al.*, 1998b, Sussman *et al.*, 2002a), which can then affect MMN generation (see Figure 2.1). It appears that the MMN process is a fairly "late" stage auditory process leading to perception, a notion that is concordant with the wealth of data suggesting that MMN elicitation is closely matched with the perception of auditory change events (e.g., Tittinen *et al.*, 1994).

7   IMPLICATIONS FOR SPEECH PERCEPTION

The notion that the segregation of sounds to sources precedes auditory event formation can be extended to include speech–processing mechanisms. The data discussed here support the view that sound elements are integrated to linguistic units (phonemes, syllables, and words) after the initial segregation or organization of the input to distinct sources. Integration of sound elements to perceptual units proceeds on the already segregated information. Speech perception, according to this model, would rely, at least in part, on the primitive ASA processes.

References

Botte, M.C., Drake, C., Brochard, R., and McAdams, S., 1997, Perceptual attenuation of nonfocused auditory streams,*Percept Psychophys.* **59**:419-425.

Bregman, A.S., 1990, *Auditory Scene Analysis*, MIT Press, MA

Brochard, R., Drake, C., Botte, M.C., and McAdams, S., 1999, Perceptual organization of complex auditory sequences: Effect of number of simultaneous sub-sequences and frequency separation, *J Exp Psychol Hum Percept Perform.* **25**:1742-1759.

Carlyon, R.P., Cusack, R., Foxton, J.M., and Robertson, I.H., 2001, Effects of attention and unilateral neglect on auditory stream segregation, *J Exp Psychol Hum Percept Perform.* **27**(1):115–127.

Fay, R.R., 2000, Spectral contrasts underlying auditory stream segregation in goldfish (Carassius auratus), *J Assoc Res Otolaryngol.* **1**:120–128.

Fishman, Y.I., Reser, D., Arezzo, J., and Steinschneider, M., 2001, Neural correlates of auditory stream segregation in primary auditory cortex of the awake monkey, *Hear Res.* **151**:167–187.

Giard, M.H., Perrin, F., Pernier, J., and Bouchet, P., 1990, Brain generators implicated in processing of auditory stimulus deviance: A topographic event-related potential study, *Psychophysiology,* **27**:627-640.

Hulse, S.H., MacDougall–Shackleton, S.A., and Wisniewski, A.B., 1997, Auditory scene analysis by songbirds: stream segregation of birdsong by European starlings (Sturnus vulgaris), *J Comp Psychol.* **111**:3–13.

Javitt, D.C., Steinschneider, M., Schroeder, C.E., Vaughan, H.G. Jr., and Arezzo, J.C., 1994, Detection of stimulus deviance within primate primary auditory cortex: intracortical mechanisms of mismatch negativity (MMN) generation, *Brain Res.* **667**:192–200.

Javitt, D.C., Steinschneider, M., Schroeder, C.E., and Arezzo, J.C., 1996, Role of cortical N–methyl–D–aspartate receptors in auditory sensory memory and mismatch–negativity generation: Implications for schizophrenia, *Proc Natl Acad Sci U S A*, **93**:11692–11967

Macken, W.J., Tremblay, S., Houghton, R.J., Nicholls, A.P., and Jones, D.M., 2003, Does auditory streaming require attention? Evidence from attentional selectivity in short–term memory, *J Exp Psychol Hum Percept Perform.* **29**(1):43–51.

Näätänen, R., Tervaniemi, M., Sussman, E., Paavilainen, and Winkler, I., 2001, Pre–attentive cognitive processing ("primitive intelligence") in the auditory cortex as revealed by the mismatch negativity (MMN), *Trends in Neuroscience,* **24,** 283–288.

Ritter, W., Sussman, E., and Molholm, S., 2000, Evidence that the mismatch negativity system works on the basis of objects, *Neuroreport.* **11**:61–63.

Sussman, E., The interaction between segregation and integration processes in auditory scene analysis. *Manuscript submitted for publication.*

Sussman E., Ceponiene, R., Shestakova, A., Näätänen, R., and Winkler, I., 2001a, Auditory stream segregation processes operate similarly in school-aged children as adults, *Hear Res.* **153**(1–2):108–114.

Sussman, E., Horváth, J., and Winkler, I., The role of attention in the formation of auditory streams. *Manuscript submitted for publication-a.*

Sussman, E., Ritter, W., and Vaughan, H.G. Jr., 1998a, Stimulus predictability and the mismatch negativity system, *Neuroreport.* **9**:4167-4170.

Sussman, E., Ritter, W., and Vaughan, H.G. Jr., 1998b, Attention affects the organization of auditory input associated with the mismatch negativity system, *Brain Res.* **789**:130–38.

Sussman, E., Ritter, W., and Vaughan, H.G. Jr., 1999, An investigation of the auditory streaming effect using event–related brain potentials, *Psychophysiology.* **36**:22–34.

Sussman, E., Sheridan, K., Kreuzer, J., and Winkler, I., 2003, Representation of the standard: stimulus context effects on the process generating the mismatch negativity component of event–related brain potentials, *Psychophysiology.* **40:** 465–471.

Sussman, E., Wang, W.J., and Bregman, A.S., Attention effects on unattended sound processes in multi-source auditory environments. *Manuscript submitted for publication-b.*

Sussman, E., and Winkler, I., 2001b, Dynamic process of sensory updating in the auditory system, *Cogn Brain Res.* **12**:431–439.

Sussman, E., Winkler, I., Huoutilainen, M., Ritter, W., and Näätänen, R., 2002a, Top–down effects on the initially stimulus–driven auditory organization, *Cogn Brain Res.* **13**:393–405.

Sussman, E., Winkler, I., Kreuzer, J., Saher, M., Näätänen, R., and Ritter, W., 2002b, Temporal integration: Intentional sound discrimination does not modify stimulus–driven processes in auditory event synthesis, *Clin Neurophysiol.* **113**:909–920.

Tiitinen, H., May, P., Reinikainen, K., and Näätänen, R., 1994, Attentive novelty detection in humans is governed by pre-attentive sensory memory, *Nature.* **372**: 90–92.

Winkler, I., Sussman, E., Tervaniemi, M., Ritter, W., Horvath, J., and Näätänen, R., 2003, Pre-attentive auditory context effects, *Cogn Affect Behav Neurosci.* **3**(1):57–77.

Yabe, H., Winkler, I., Czigler, I., Koyama, S., Kakigi, R., Sutoh, T., and Kaneko, S., 2001, Organizing sound sequences in the human brain: the interplay of auditory streaming and temporal integration. *Brain Res.* **897**:222–227.

Chapter 3

Speech separation:
Further Insights from Recordings of Event–related Brain Potentials in Humans

Claude Alain

Rotman Research Institute, Baycrest Centre for Geriatric Care and Department of Psychology, University of Toronto
calain@rotman-baycrest.on.ca

1   INTRODUCTION

Sounds are created by a wide range of acoustic sources, such as several people talking during a cocktail party. The typical source generates complex acoustic energy that has many frequency components. In a quiet environment, it is usually easy to understand what a person is saying. In many listening situations however, different acoustic sources are active at the same time, and only the sum of those spectra will reach the listener's ears. Therefore, for individual sound patterns to be recognized – such as those arriving from a particular human voice among a mixture of many – the incoming auditory information must be partitioned, and the correct subset of elements must be allocated to individual sounds so that a veridical description may be formed for each. This is a complicated task because each ear has access only to a single pressure wave that is the sum of the pressure waves from all individual sound sources. The process by which we decompose this complex acoustic wave has been termed auditory scene analysis (Bregman, 1990), and it involves perceptually organizing our environment along at least two axes: time and frequency. Organization along the time axis entails the sequential grouping of acoustic data over several seconds, whereas processing along the frequency axis involves the segregation of simultaneous sound sources according to their different frequencies and harmonic relations.

Generally speaking, auditory scene analysis theory seeks to explain how the auditory system assigns acoustic elements to different sound sources. Bregman (1990) proposed a general theory of auditory scene analysis based primarily on the Gestalt laws of organization (Koffka, 1935). In this framework, auditory scene analysis is divided into two classes of processes, dealing with the perceptual organization of simultaneously (i.e., concurrent) and sequentially occurring acoustic elements, respectively. These processes are responsible for grouping and parsing components of the acoustic mixture to construct perceptual representations of sound sources, or 'auditory objects', according to principles such as physical similarity, temporal proximity, and good continuation. For example, sounds are more likely to be assigned to separate representations if they differ widely in frequency, intensity, and/or spatial location. In contrast, sound components that are harmonically related, or that rise and fall in intensity together, are more likely to be perceptually grouped and assigned to a single source. Many of these processes are considered automatic or 'primitive' since they can be found in infants (Winkler *et al.*, 2003) and animals such as birds (Hulse *et al.*, 1997, MacDougall-Shackleton *et al.*, 1998) and monkeys (Fishman *et al.*, 2001). The outcome of this pre-attentive analysis may then be subjected to a more detailed analysis by controlled (i.e., top–down) processes. While the pre-attentive processes group sounds based on physical similarity, controlled schema–driven processes apply prior knowledge to constrain the auditory scene, leading to perceptions that are consistent with previous experience. As such, schema–driven processes depend on both the representations of previous auditory experience acquired through learning, and a comparison of the incoming sounds with those representations. The use of prior knowledge is particularly useful during adverse listening situations, such as the cocktail party situation described above. In an analogous laboratory situation, a sentence's final word embedded in noise is more easily detected when it is contextually predictable than when it is unpredictable (e.g., "His plan meant taking a big risk." as opposed to "Jane was thinking about the oath.") (Pichora-Fuller *et al.*, 1995). Thus, schema–driven processes provide a mechanism for resolving perceptual ambiguity in complex listening situations.

Deficits in listeners' aptitude to perceptually organize auditory input could have dramatic consequences on the perception and identification of complex auditory signals such as speech and music. For example, impairment in the ability to adequately separate the spectral components of sequentially and/or simultaneously occurring sounds may contribute to speech perception problems often observed in older adults (Alain *et al.*, 2001, Divenyi and Haupt, 1997a, 1997b, 1997c, Grimault *et al.*, 2001) and in individuals with dyslexia (Helenius *et al.*, 1999, Sutter *et al.*, 2000). Hence, understanding how the brain solves complex auditory scenes that unfold over time is a major goal for both psychological and physiological sciences. Although auditory scene analysis has been investigated extensively for almost 30 years and there have been several attempts to use computer models to simulate or reproduce the phenomena found in an increasingly extensive literature, major gaps remain between psychoacoustic research and neurophysiological research. This chapter is an attempt to bridge together findings from different disciplines in hearing sciences. I briefly review neurophysiological studies in both animals and humans that provide important insights into the neural basis of auditory scene analysis in general and in speech separation in particular. I then present some preliminary data that illustrate a new neural correlate of speech segregation.

2 BIOLOGICAL FOUNDATION OF AUDITORY SCENE ANALYSIS

The neurobiological foundation of auditory scene analysis has received considerable attention over the last decade. Evidence from single cell recordings shows that frequency periodicity, upon which concurrent sound segregation is partly based, is reflected within the patterns of afferent spike trains (Bodnar and Bass, 1999, Cariani and Delgutte, 1996a, 1996b, Keilson *et al.*, 1997, Palmer, 1990, Sinex *et al.*, 2002). Multi-unit recordings in non-human primates have also revealed a distinct pattern of neural activity in primary auditory cortex associated with conditions that promote sequential auditory stream formation (Fishman *et al.*, 2001). This suggests that both spectral and temporal transitions between successive stimuli are represented within the primary auditory cortex. Although these results suggest early bottom–up (stimulus–driven) processes in auditory scene analysis, representations of incoming acoustic information in the ascending auditory pathway are probably not sufficient for the detection and identification of different sound objects. Griffiths and Warren (Griffiths and Warren, 2002) argue that the discrimination and identification of auditory objects requires additional computations that follow the initial processing in the ascending pathway and primary auditory cortex, suggesting that these computations might be carried out in the planum temporale.

In addition, it has been proposed that identifying the content (what) and the location (where) of sound in the environment may be functionally segregated in a manner analogous to the ventral (what) and dorsal (where) pathways in the visual modality (Rauschecker, 1997, Rauschecker and Tian, 2000). This idea has received considerable support from neuroanatomical (Romanski and Goldman-Rakic, 2002, Romanski *et al.*, 1999) and neurophysiological studies in nonhuman primates (Rauschecker, 1997, 1998, Rauschecker and Tian, 2000, Tian *et al.*, 2001), as well as lesion (Adriani *et al.*, 2003, Clarke *et al.*, 1996, Clarke *et al.*, 2000, Clarke *et al.*, 2002) and neuroimaging (Alain *et al.*, 2001, Maeder *et al.*, 2001) studies in humans. Results from a recent meta-analysis of studies employing either positron emission tomography (PET) or functional magnetic resonance imaging (fMRI) provide further support for a dual pathway model of auditory attribute processing (Arnott, Binns, Grady, and Alain, In press). In that analysis, most studies employing tasks that required judgments about sound location reported activation in inferior parietal cortex as well as superior frontal gyrus. In comparison, most studies involving the processing of sound identity (e.g., phoneme discrimination, pitch discrimination etc.) reported activation in the anterior portion of the temporal lobe and the inferior frontal gyrus. While neuroimaging studies in humans have identified a number of regions that may contribute to auditory scene analysis, little is known about the time course of these neural events and how they relate to phenomenological experience.

2.1   Event–Related Brain Potential Studies of Auditory Scene Analysis

Recording human event–related brain potentials (ERPs) is a powerful measure for examining how auditory scene analysis unfolds over time because this technique allows for the examination of neural activity within hundreds of milliseconds after the presentation of a sound either within or outside the focus of attention. Consequently, ERPs can be used to evaluate the effect of variables such as attention (Alain and Izenberg, 2003, Hillyard *et al.*, 1973, Martin *et al.*, 1997) and learning (Shahin *et al.*, 2003, Tremblay *et al.*, 1997, Wayman *et al.*, 1992). Basically speaking, ERPs reflect the synchronous activity from large neuronal ensembles that are time–locked to sensory or cognitive events. Auditory ERPs therefore represent the journey of acoustic information as it ascends the auditory system from the cochlea through the brainstem to the primary auditory cortex and on to higher auditory cortical areas. Brainstem auditory evoked potentials occur between 1 and 10 ms after stimulus onset. Middle–latency evoked potentials arise between 10 and 50 ms after stimulus presentation and are thought to reflect the activity of the primary auditory cortex. Long–latency evoked potentials take place after 50 ms and include the P1, N1, and P2 wave. The N1–P2 complex is thought to be of particular theoretical interest since it is related to signal detection, and is present only when a transient auditory stimulus is audible. However, the conscious identification of an auditory event is often associated with an additional late positive wave peaking between 250 and 600 ms post–stimulus, referred to as the P300 or P3b (Hillyard *et al.*, 1971, Martin *et al.*, 1997, Parasuraman and Beatty, 1980, Parasuraman *et al.*, 1982).

2.2   The mistuned harmonic paradigm

One way of investigating the neural underpinnings of concurrent sound segregation is by means of the mistuned harmonic paradigm. In such a paradigm, the listener is usually presented with two successive stimuli, one comprised of entirely harmonic components, the other with a mistuned harmonic. The task of the listener is to indicate which one of the two stimuli contains the mistuned harmonic. Several factors influence the perception of the mistuned harmonic including degree of inharmonicity, harmonic number

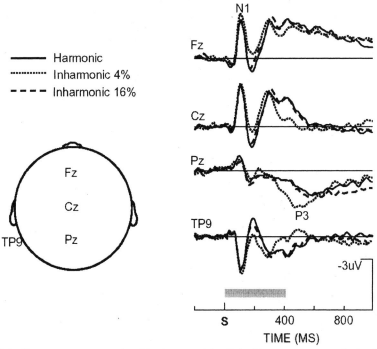

Figure 3.1. Group mean event–related brain potentials elicited by complex sounds that had all harmonics in tune or had the second harmonic shifted upward by 4% or 16% of its original value. The gray rectangle illustrates sound duration. S refers to stimulus onset. Negativity is plotted upward.

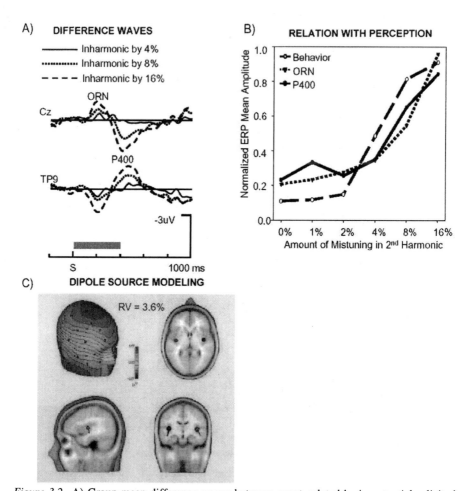

Figure 3.2. A) Group mean difference waves between event–related brain potentials elicited by complex sounds that have all harmonics in tune and by sounds with the second harmonic shifted upward by 4, 8, or 16% of its original value. The gray rectangle illustrates sound duration. **S** refers to sound onset. Each tick marks a 200-ms interval. B) Relation between participants' likelihood to report perceiving two auditory objects and object–related negativity (ORN) and P400 amplitudes. Note that the amplitude of the neuroelectric responses was normalized so that brain responses and performance could be displayed on the same scale. C) Dipole source modeling of the ORN. RV = residual variance of the model. In the present context, the low RV indicates a very good fit of the data. The darker grey in the contour map (upper right) indicates greater negativity.

and sound duration (Hartmann *et al.*, 1990, Lin and Hartmann, 1998, Moore *et al.*, 1985).

In a series of experiments, Alain *et al.* (2001) measured ERPs to loudness–matched complex sounds composed of either all tuned harmonics (or "partials") or multiple tuned and one mistuned harmonic. On each trial, participants indicated whether they heard a single sound (i.e., a regular "buzz" at a pitch equivalent to the fundamental) or two sounds (i.e., the buzz plus a separate sound with a pure–tone quality at the frequency of the mistuned harmonic). In this context, the mistuned harmonic might be analogous to the presence of a secondary voice related to another fundamental frequency. When the harmonic in the complex sound was mistuned from its original value by more than 4%, listeners heard it as a separate tone.

This perception of multiple, simultaneous auditory objects was accompanied by a negative displacement in the waveform that we named the "object–related negativity" (ORN). The ORN overlaps in time with the N1 and P2 deflections (see Figures 3.1 and 3.2). Moreover, the ORN amplitude correlated with perceptual judgment, being greater when participants reported hearing two distinct perceptual objects (the buzz and the tone). It is observed in school-aged children (Alain *et al.*, 2003) and can be recorded for sounds that are segregated based on location rather than harmonic cues (Johnson *et al.*, 2003).

The ORN can be recorded for stimuli that are unattended, such as when participants ignore the stimuli and read a book of their choice (Alain *et al.*, 2001) or watch a subtitled movie (Alain *et al.*, 2002). The fact that the ORN can be recorded even when participants are not attending to the stimuli is consistent with the proposal that concurrent sound segregation may occur independently of a listener's attention. However, these findings should be interpreted with caution because, in these studies, listener attention was not well controlled. For example, participants were required to take part in a primary visual activity and to ignore the auditory stimuli, but little effort was made to ensure that they complied with these instructions. Since there was no objective measure of attention, the possibility that the participant's attention may have occasionally wandered to the auditory stimuli cannot be ruled out. In a more recent study, we examined the effects of attention on the neural processes underlying concurrent sound segregation through the manipulation of auditory task load (Alain and Izenberg, 2003). Participants were asked to focus their attention on tuned and mistuned stimuli presented to one ear (e.g., left) and to ignore similar stimuli presented to the other ear (e.g., right). For both tuned and mistuned sounds, long (standard) and shorter (deviant) duration stimuli were presented to both ears. Auditory task load was manipulated by varying task instructions. In the easier condition, participants were asked to press a button for deviant sounds at the attended location, irrespective of tuning. In the harder condition, participants were further asked to identify whether the targets were tuned or mistuned. Participants were faster in detecting targets defined by duration only than by both duration <u>and</u> tuning. More importantly, at the unattended location mistuned stimuli generated an ORN whose amplitude was not affected by task difficulty. These results provide strong support for the proposal that concurrent sound segregation can take place independently of listener attention.

The ORN amplitude is usually largest at central and frontocentral sites and inverts polarity at the mastoid sites, consistent with generators located in the supratemporal plane within the Sylvian fissure. Dipole source modeling suggests that the ORN sources are inferior and medial to N1 sources (Alain *et al.*, 2001), indicating that neurons activated by co-occurring auditory stimuli are different from those activated by stimulus onset. In an effort to further identify the neural substrates of concurrent sound segregation, we measured middle latency auditory evoked responses (i.e., Na and Pa) which are generated primarily within the primary auditory cortex (Liegeois-Chauvel *et al.*, 1995). The Pa wave at about 30 ms was significantly larger when the third harmonic of a 200 Hz complex sound was mistuned by 16% of its original value (Dyson and Alain, 2004). In a manner similar to the ORN, the enhanced Pa amplitude was also paralleled by an increased likelihood of participants reporting the presence of multiple, concurrent auditory objects. These results are consistent with an early stage of auditory scene analysis in which acoustic properties such as mistuning act as pre-attentive segregation cues that can subsequently lead to the perception of multiple auditory objects. It also suggests that the primary auditory cortex (the main source of the Pa wave) represents inharmonicity and therefore may play an important role in the initial stage of concurrent sound segregation.

Distinguishing simultaneous auditory objects is also accompanied by a late positive wave (P400), which has a widespread scalp distribution. Like the ORN, its amplitude is correlated with perceptual judgment, being larger when participants perceive the mistuned harmonic as a separate tone. However, in contrast with the ORN, this component is present only when participants are required to attend to the stimuli and respond whether they heard one or two auditory stimuli. Using sounds of various duration, Alain *et al.* (2002) showed that the P400 can be dissociated from motor processes. Thus, whereas the ORN appears to be more associated with automatic processing, the P400 seems to be more related to controlled processes.

This sequence of neural events underlying the perception of simultaneous auditory objects is consistent with Bregman's account of auditory scene analysis. Specifically, harmonically related partials are grouped together into one entity, while the partial that is sufficiently mistuned stands out of the complex as a separate object. The ORN thus may be viewed as indexing the automatic detection of the mistuned harmonic from a prediction based upon the expected harmonics of the incoming stimulus. This is consistent with psychophysical data, which argue for a pattern–matching process that attempts to adjust a harmonic template (or sieve), defined by a fundamental frequency, to fit the spectral pattern (Goldstone, 1978, Hartmann, 1996, Lin and Hartmann, 1998). When a harmonic is mistuned by a sufficient amount, a discrepancy occurs between the perceived frequency and that frequency expected on the basis of the template. The purpose of this pattern–matching process would be to signal to higher auditory centers that more than one auditory object might be simultaneously present in the environment. Within this context, the P400 component may index a schema–driven process involved in scene analysis. This proposal is supported by findings showing that the P400 amplitude, like perception of the mistuned harmonic as a separate object, decreases when stimuli are presented frequently (Alain *et al.*, 2001).

The ORN shows some similarities in latency and amplitude distribution with another ERP component called the mismatch negativity (MMN). The MMN is elicited by the occurrence of rare deviant sounds embedded in a sequence of homogenous standard stimuli (Näätänen, 1992, Picton *et al.*, 2000). Like the ORN, the MMN has a frontocentral distribution and its latency peaks at about 150 ms after the onset of deviation. Both ORN and MMN to acoustic stimuli can be recorded while listeners are reading or watching a video and therefore are thought to index bottom–up processing of auditory scene analysis. Despite the similarities, these two ERP components differ in three important ways. One of the most crucial differences between these two components is that while MMN is highly sensitive to the perceptual context, the ORN is not. That is, the MMN is elicited only by <u>rare</u> deviant stimuli whereas the ORN is elicited by mistuned stimuli regardless of whether they are presented occasionally or frequently (Alain *et al.*, 2001). Thus, the MMN reflects a mismatch between the incoming auditory stimulus and what is expected based on recently occurring stimuli, whereas the ORN indexes a discrepancy between the mistuned harmonic and what is expected on the basis of the current stimulus. Thus, the MMN can be viewed as an index of sequential integration because its elicitation depends on the extraction of regularities over several seconds. In comparison, the ORN is thought to index concurrent sound segregation and depends on an online spectral analysis of the incoming acoustic waveform.

The ORN and MMN can also be distinguished based on their scalp distribution. Scalp distributions and dipole source modeling are important critera in identifying and distinguishing between ERP components. Figure 3.3 shows the ORN and MMN amplitude distribution in the same group of participants. The MMN was elicited by shorter duration stimuli and showed a more frontally distributed response compared with the ORN elicited by complex sounds with the third harmonic mistuned by 16%. Differences in amplitude distribution indicate that different neural networks are responsible for concurrent and sequential sound segregation. Lastly, the ORN and MMN appear to be differentially sensitive to attentional manipulation. In a recent study, we showed that the ORN was little affected by attention load whereas the MMN was reduced in amplitude when the auditory task was more demanding (Alain and Izenberg, 2003). The effects of selective attention on MMN amplitude have been reported in many studies for a variety of deviant types (Alain and Arnott, 2000). This difference in attentional sensitivity may be related to the memory system upon which the grouping processes depend. That is, sequential integration depends on the integration of acoustic information over several seconds while concurrent sound segregation depends on the integration of acoustic information within hundreds of milliseconds.

3   NEURAL CORRELATES OF SPEECH SEPARATION

Early classical work with dichotic listening has shown that a listener's ability to understand speech depends on the acoustical factors that promote the segregation of co–occurring speech stimuli into distinct sound objects. For example, an increase in spatial separation between two concurrent messages improves performance in identifying the task–relevant message (Bronkhorst and Plomp, 1988, 1992, Spieth, Curtis, and Webster, 1954, Treisman, 1964).

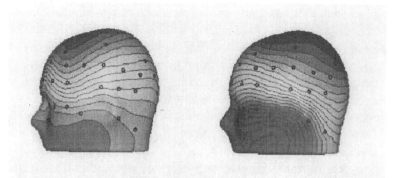

Figure 3.3. Contour maps showing the amplitude distribution of the MMN and ORN within the same group of participants. The MMN was elicited by short duration deviant sounds whereas the ORN was recorded from complex sounds with the third harmonic mistuned by 16% of its original value. The darker grey indicates greater negativity. Note the difference in the orientation of the contour lines between the two maps over the temporal lobe.

Similarly, increasing the distance between the frequency bands of the two messages improves an individual's ability to attend to either of the messages selectively (Spieth *et al.*, 1954). Psychophysical studies have also shown that when listeners are presented with two different vowels simultaneously, the identification rate improves with increasing separation between the fundamental frequencies (f_0) of the two vowels (Assmann and Summerfield, 1990, 1994, Chalikia and Bregman, 1989, Culling and Darwin, 1993). These behavioural studies highlight the importance of f_0 differences in the perceptual separation of competing voices.

We recently investigated the time course of neural activity associated with concurrent vowel segregation (Alain *et al.*, Submitted; Reinke *et al.*, 2003), using a paradigm similar to that of Assmann and Summerfield (1994). Participants were informed that on each trial two phonetically different vowels would be presented simultaneously and they were asked to identify both by pressing the corresponding keys on the keyboard as the difference in f_0 varied from trial to trial. As previously reported in the behavioral literature (Assmann and Summerfield, 1990, 1994, Chalikia and Bregman, 1989), we found that a listener's ability to identify both vowels improved by increasing the difference in f_0 between the two

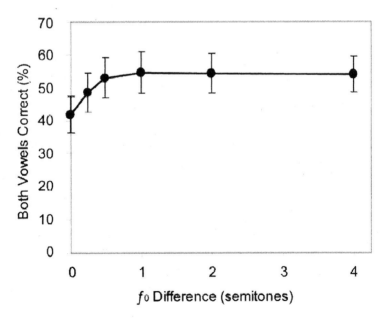

Figure 3.4. Proportion of correctly identified vowels as a function of $f0$.

vowels (Figure 3.4). We also found that the listener's ability to identify two concurrent vowels improved with training (Reinke *et al.*, 2003), and that improvement was associated with decreased N1 and P2 latencies and an enhancement in P2 amplitude. These learning–related changes in sensory evoked responses may reflect functional and/or structural changes in auditory cortices that could reflect an increase in listener expertise with such stimuli.

The correct identification of concurrent vowels depends on a listener's ability to detect the presence of two signals, identify these individual signals and to initiate the appropriate response set. With respect to these processes, we found two ERP signatures that may underlie the detection and identification of concurrent vowels, respectively (Figure 3.5). The first ERP modulation was a negative wave that was superimposed on the N1 and P2 waves, and peaked around 140 ms after sound onset. This component was maximal over midline central electrodes and showed similarities in latency and amplitude distribution with the ORN. As with the ORN, the amplitude of this component was related to the detection of the discrepancy between f_0's, signaling to higher auditory centers that two sound sources were present. In the case where f_0 is zero, some of the harmonics may activate similar populations of neurons responding to the temporal characteristics of the steady state vowel and/or the "place" on the cortex (Ohl and Scheich, 1997), whereas different populations may be activated as f_0 becomes larger.

The second ERP component associated with concurrent vowel segregation was a negative wave that peaked at about 250 ms after sound onset and was larger over the right and central regions of the scalp. As mistuned harmonics do not generate this late modulation, it was likely related to the identification and categorization that followed the automatic detection of the two constituents of the double vowel stimuli. The first negative modulation was present whether or not the participants were involved in identifying the two constituents of the double–vowel stimulus. This suggests that detecting whether two different vowels were present in the mixture occurred automatically. The second negative peak was present only when listeners were required to make a perceptual decision. This component may index a matching process between the incoming signal and the stored representation of the vowels in working memory. Given that vowels are over–learned, the second modulation may also reflect the influence of schema–driven processes in vowel identification.

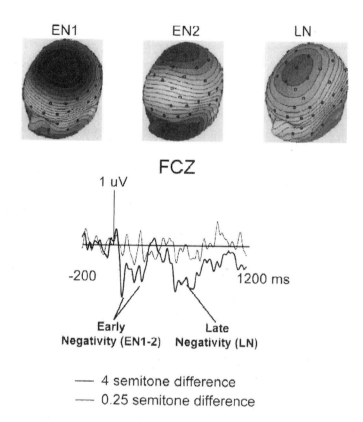

Figure 3.5. Group mean difference waves between event–related brain potentials elicited by two vowels sharing the same fundamental frequency or those recorded when the two vowels were separated by .25 or 4 semitones. Contour maps illustrate the amplitude distribution of the early (EN1 and EN2) and late (LN) negativity. The EN1, EN2, and LN peak around 150, 250, and 650 ms after sound onset. Negativity is plotted downward. FCZ = FrontoCentral electrode at the midline. The darker grey in the contour maps indicates greater negativity.

4   SUMMARY AND CONCLUSION

A fundamental problem faced by the human auditory system is the segregation of concurrent speech signals. To discriminate between individual voices, listeners must extract information from the composite acoustic wave reflecting summed activation from all the simultaneously active voices. In this chapter, we showed that an observer's ability to identify two different vowels presented simultaneously improved by increasing the fundamental frequency (f_0) separation between the vowels. Recording of brain event–related potentials for stimuli presented during attend and ignore conditions revealed activity (i.e., object–related negativity, ORN) between 130 and 170 ms after sound onset that coded f_0 differences. Another, more right–lateralized negative wave maximal at 250 ms accompanied the improvement in vowel identification. This sequence of neural events is consistent with a multistage model of auditory scene analysis in which the spectral pattern of each vowel constituent is first extracted and then matched against representations of those vowels in working memory. We found a different pattern of neural activity when individuals were required to identify concurrent vowels than when they were asked whether one or two auditory events were present in a mixture using the mistuned harmonic paradigm (Alain *et al.*, 2001, Alain *et al.*, 2002). These differences may be related to observer expertise with the speech material, i.e., the processing of vowels may automatically engage both low–level and schema–driven processes. Our findings highlight the need to further investigate and test models of auditory perception and cognition using more ecologically valid stimuli.

5   ACKNOWLEDGMENTS

Supported by the CIHR, NSERC, and The Hearing Foundation of Canada. I wish to thank B. Dyson and J. Snyder for their comments on the manuscript and valuable discussion. I am particularly indebted to Peter Assmann and Quentin Summerfield for providing the vowel stimuli. This work was supported by a grant from the Canadian Institute for Health Research.

References

Adriani, M., Maeder, P., Meuli, R., Thiran, A.B., Frischknecht, R., Villemure, J.G., *et al.*, 2003, Sound recognition and localization in man: specialized cortical networks and effects of acute circumscribed lesions. *Exp Brain Res,* **153(4)**, 591–604.

Alain, C. and Arnott, S.R., 2000, Selectively attending to auditory objects. *Front Biosci,* **5**, D202–212.

Alain, C., Arnott, S.R., Hevenor, S., Graham, S., and Grady, C.L., 2001, "What" and "where" in the human auditory system. *Proc Natl Acad Sci U S A,* **98(21)**, 12301–12306.

Alain, C., Arnott, S.R., and Picton, T.W., 2001, Bottom–up and top–down influences on auditory scene analysis: evidence from event–related brain potentials. *J Exp Psychol Hum Percept Perform,* **27(5)**, 1072–1089.

Alain, C. and Izenberg, A., 2003, Effects of Attentional Load on Auditory Scene Analysis. *J Cogn Neurosci,* **15(7)**, 1063–1073.

Alain, C., McDonald, K.L., Ostroff, J.M., and Schneider, B., 2001, Age–related changes in detecting a mistuned harmonic. *J Acoust Soc Am*, **109(5 Pt 1)**, 2211–2216.

Alain, C., Reinke, K.S., He, Y., Wang, C., and Lobaugh, N. (Submitted). Hearing two things at once: Neurophysiological indices of speech segregation and identification. *Journal of Cognitive Neuroscience.*

Alain, C., Schuler, B.M., and McDonald, K.L., 2002, Neural activity associated with distinguishing concurrent auditory objects. *J Acoust Soc Am*, **111(2)**, 990–995.

Alain, C., Theunissen, E.L., Chevalier, H., Batty, M., and Taylor, M.J., 2003, Developmental changes in distinguishing concurrent auditory objects. *Brain Res Cogn Brain Res,* **16(2)**, 210–218.

Arnott, S.R., Binns, M.A., Grady, C.L., and Alain, C. (In press). Assessing the auditory dual–pathway model in humans. *NeuroImage.*

Assmann, P. and Summerfield, Q., 1990, Modeling the perception of concurrent vowels: Vowels with different fundamental frequencies. *J Acoust Soc Am*, **88(2)**, 68097.

Assmann, P. and Summerfield, Q., 1994, The contribution of waveform interactions to the perception of concurrent vowels. *J Acoust Soc Am*, **95(1)**, 471–484.

Bodnar, D.A. and Bass, A.H., 1999, Midbrain combinatorial code for temporal and spectral information in concurrent acoustic signals. *J Neurophysiol,* **81(2)**, 552–563.

Bregman, A.S., 1990, *Auditory Scene Analysis: The Perceptual Organization of Sounds.* London, England: The MIT Press.

Bronkhorst, A.W. and Plomp, R., 1988, The effect of head–induced interaural time and level differences on speech intelligibility in noise. *J Acoust Soc Am*, **83(4)**, 1508–1516.

Bronkhorst, A.W. and Plomp, R., 1992, Effect of multiple speechlike maskers on binaural speech recognition in normal and impaired hearing. *J Acoust Soc Am*, **92(6)**, 3132–3139.

Cariani, P.A. and Delgutte, B., 1996a, Neural correlates of the pitch of complex tones. I. Pitch and pitch salience. *J Neurophysiol,* **76(3)**, 1698–1716.

Cariani, P.A. and Delgutte, B., 1996b, Neural correlates of the pitch of complex tones. II. Pitch shift, pitch ambiguity, phase invariance, pitch circularity, rate pitch, and the dominance region for pitch. *J Neurophysiol,* **76(3)**, 1717–1734.

Chalikia, M.H. and Bregman, A.S., 1989, The perceptual segregation of simultaneous auditory signals: pulse train segregation and vowel segregation. *Percept Psychophys,* **46(5)**, 487–496.

Clarke, S., Bellmann, A., De Ribaupierre, F., and Assal, G., 1996, Non–verbal auditory recognition in normal subjects and brain–damaged patients: evidence for parallel processing. *Neuropsychologia,* **34(6)**, 587–603.

Clarke, S., Bellmann, A., Meuli, R.A., Assal, G., and Steck, A.J., 2000, Auditory agnosia and auditory spatial deficits following left hemispheric lesions: evidence for distinct processing pathways. *Neuropsychologia,* **38(6)**, 797–807.

Clarke, S., Bellmann Thiran, A., Maeder, P., Adriani, M., Vernet, O., Regli, L., *et al.*, 2002, What and where in human audition: selective deficits following focal hemispheric lesions. *Exp Brain Res,* **147(1)**, 8–15.

Culling, J.F. and Darwin, C.J., 1993, Perceptual separation of simultaneous vowels: Within and across–formant grouping by F_0. *J Acoust Soc Am*, **93(6)**, 3454–3467.

Divenyi, P.L. and Haupt, K.M., 1997a, Audiological correlates of speech understanding deficits in elderly listeners with mild–to–moderate hearing loss. I. Age and lateral asymmetry effects. *Ear Hear,* **18(1)**, 42–61.

Divenyi, P. L. and Haupt, K.M., 1997b, Audiological correlates of speech understanding deficits in elderly listeners with mild–to–moderate hearing loss. II. Correlation analysis. *Ear Hear,* **18(2)**, 100–113.

Divenyi, P.L. and Haupt, K.M., 1997c, Audiological correlates of speech understanding deficits in elderly listeners with mild–to–moderate hearing loss. III. Factor representation. *Ear Hear,* **18(3)**, 189–201.

Dyson, B. and Alain, C., 2004, Representation of sound object in primary auditory cortex. *J Acoust Soc Am,* **115**, 280–288.

Fishman, Y.I., Reser, D.H., Arezzo, J.C., and Steinschneider, M., 2001, Neural correlates of auditory stream segregation in primary auditory cortex of the awake monkey. *Hear Res,* **151(1–2)**, 167–187.

Goldstein, J.L., 1978, Mechanisms of signal analysis and pattern perception in periodicity pitch. *Audiology,* **17(5)**, 421–445.

Griffiths, T.D. and Warren, J.D., 2002, The planum temporale as a computational hub. *Trends Neurosci,* **25(7)**, 348–353.

Grimault, N., Micheyl, C., Carlyon, R.P., Arthaud, P., and Collet, L., 2001, Perceptual auditory stream segregation of sequences of complex sounds in subjects with normal and impaired hearing. *Br J Audiol,* **35(3)**, 173–182.

Hartmann, W.M., 1996, Pitch, periodicity, and auditory organization. *J Acoust Soc Am,* **100(6)**, 3491–3502.

Hartmann, W.M., McAdams, S., and Smith, B.K., 1990, Hearing a mistuned harmonic in an otherwise periodic complex tone. *J Acoust Soc Am,* **88(4)**, 1712–1724.

Helenius, P., Uutela, K., and Hari, R., 1999, Auditory stream segregation in dyslexic adults. *Brain,* **122(Pt 5)**, 907–913.

Hillyard, S.A., Hink, R.F., Schwent, V.L., and Picton, T.W., 1973, Electrical signs of selective attention in the human brain. *Science,* **182(108)**, 177–180.

Hillyard, S.A., Squires, K.C., Bauer, J.W., and Lindsay, P.H., 1971, Evoked potential correlates of auditory signal detection. *Science,* **172(990)**, 1357–1360.

Hulse, S.H., MacDougall-Shackleton, S.A., and Wisniewski, A.B., 1997, Auditory scene analysis by songbirds: stream segregation of birdsong by European starlings (Sturnus vulgaris). *J Comp Psychol,* **111(1)**, 3–13.

Johnson, B.W., Hautus, M., and Clapp, W.C., 2003, Neural activity associated with binaural processes for the perceptual segregation of pitch. *Clin Neurophysiol,* **114(12)**, 2245–2250.

Keilson, S.E., Richards, V.M., Wyman, B.T., and Young, E.D., 1997, The representation of concurrent vowels in the cat anesthetized ventral cochlear nucleus: evidence for a periodicity–tagged spectral representation. *J Acoust Soc Am,* **102(2 Pt 1)**, 1056–1071.

Koffka, K., 1935, *Principles of Gestalt Psychology.* New York: Harcout, Brace, and World.

Liegeois-Chauvel, C., Laguitton, V., Badier, J.M., Schwartz, D., and Chauvel, P., 1995, [Cortical mechanisms of auditive perception in man: contribution of cerebral potentials and evoked magnetic fields by auditive stimulations]. *Rev Neurol (Paris),* **151(8–9)**, 495–504.

Lin, J.Y. and Hartmann, W.M., 1998, The pitch of a mistuned harmonic: evidence for a template model. *J Acoust Soc Am,* **103(5 Pt 1)**, 2608–2617.

MacDougall-Shackleton, S.A., Hulse, S.H., Gentner, T.Q., and White, W., 1998, Auditory scene analysis by European starlings (Sturnus vulgaris): perceptual segregation of tone sequences. *J Acoust Soc Am,* **103(6)**, 3581–3587.

Maeder, P.P., Meuli, R.A., Adriani, M., Bellmann, A., Fornari, E., Thiran, J.P., *et al.*, 2001, Distinct pathways involved in sound recognition and localization: a human fMRI study. *Neuroimage,* **14(4)**, 802–816.

Martin, B.A., Sigal, A., Kurtzberg, D., and Stapells, D. R., 1997, The effects of decreased audibility produced by high–pass noise masking on cortical event–related potentials to speech sounds /ba and /da. *J Acoust Soc Am,* **101(3)**, 1585–1599.

Moore, B.C., Peters, R.W., and Glasberg, B.R., 1985, Thresholds for the detection of inharmonicity in complex tones. *J Acoust Soc Am,* **77(5)**, 1861–1867.

Näätänen, R., 1992, *Attention and brain function.* Hillsdale: Erlbaum.

Ohl, F.W. and Scheich, H., 1997, Orderly cortical representation of vowels based on formant interaction. *Proc Natl Acad Sci U S A,* **94(17)**, 9440–9444.

Palmer, A.R., 1990, The representation of the spectra and fundamental frequencies of steady-state single– and double–vowel sounds in the temporal discharge patterns of guinea pig cochlear–nerve fibers. *J Acoust Soc Am,* **88(3)**, 1412–1426.

Parasuraman, R. and Beatty, J., 1980, Brain events underlying detection and recognition of weak sensory signals. *Science,* **210(4465)**, 80–83.

Parasuraman, R., Richer, F., and Beatty, J., 1982, Detection and recognition: Concurrent processes in perception. *Percept Psychophys,* **31(1)**, 1–12.

Pichora-Fuller, M.K., Schneider, B.A., and Daneman, M., 1995, How young and old adults listen to and remember speech in noise. *J Acoust Soc Am,* **97(1)**, 593–608.

Picton, T.W., Alain, C., Otten, L., Ritter, W., and Achim, A., 2000, Mismatch negativity: Different water in the same river. *Audiol Neurootol,* **5(3–4)**, 111–139.

Rauschecker, J.P., 1997, Processing of complex sounds in the auditory cortex of cat, monkey, and man. *Acta Otolaryngol Suppl,* **532**, 34–38.

Rauschecker, J.P., 1998, Parallel processing in the auditory cortex of primates. *Audiol Neurootol,* **3(2–3)**, 86–103.

Rauschecker, J.P. and Tian, B., 2000, Mechanisms and streams for processing of ",what" and "where" in auditory cortex. *Proc Natl Acad Sci U.S.A.,* **97(22)**, 11800–11806.

Reinke, K.S., He, Y., Wang, C., and Alain, C., 2003, Perceptual learning modulates sensory evoked response during vowel segregation. *Brain Res Cogn Brain Res,* **17(3)**, 781–791.

Romanski, L.M. and Goldman-Rakic, P.S., 2002, An auditory domain in primate prefrontal cortex. *Nat Neurosci,* **5(1)**, 15–16.

Romanski, L.M., Tian, B., Fritz, J., Mishkin, M., Goldman-Rakic, P.S., and Rauschecker, J.P., 1999, Dual streams of auditory afferents target multiple domains in the primate prefrontal cortex. *Nat Neurosci,* **2(12)**, 1131–1136.

Shahin, A., Bosnyak, D.J., Trainor, L.J., and Roberts, L.E., 2003, Enhancement of neuroplastic P2 and N1c auditory evoked potentials in musicians. *J Neurosci,* **23(13)**, 5545–5552.

Sinex, D.G., Henderson Sabes, J., and Li, H., 2002, Responses of inferior colliculus neurons to harmonic and mistuned complex tones. *Hear Res,* **168(1–2)**, 150–162.

Spieth, W., Curtis, J.F., and Webster, J.C., 1954, Responding to one of two simultaneous messages. *J Acoust Soc Am,* **26(3)**, 391–396.

Sutter, M.L., Petkov, C., Baynes, K., and O'Connor, K.N., 2000, Auditory scene analysis in dyslexics. *Neuroreport,* **11(9)**, 1967–1971.

Tian, B., Reser, D., Durham, A., Kustov, A., and Rauschecker, J.P., 2001, Functional specialization in rhesus monkey auditory cortex. *Science,* **292(5515)**, 290–293.

Treisman, A., 1964, The effect of irrelevant material on the efficiency of selective listening. *The American Journal of Psychology,* **77**, 533–546.

Tremblay, K., Kraus, N., Carrell, T.D., and McGee, T., 1997, Central auditory system plasticity: generalization to novel stimuli following listening training. *J Acoust Soc Am,* **102(6)**, 3762–3773.

Wayman, J.W., Frisina, R.D., Walton, J.P., Hantz, E.C., and Crummer, G.C., 1992, Effects of musical training and absolute pitch ability on event–related activity in response to sine tones. *J Acoust Soc Am,* **91(6)**, 3527–3531.

Winkler, I., Kushnerenko, E., Horvath, J., Ceponiene, R., Fellman, V., Huotilainen, M., *et al.,* 2003, Newborn infants can organize the auditory world. *Proc Natl Acad Sci U S A,* **100(20)**, 11812–11815.

Chapter 4

Recurrent Timing Nets for F0–based Speaker Separation

Peter Cariani

Eaton Peabody Laboratory of Auditory Physiology
Massachusetts Eye & Ear Infirmary, Boston, Massachusetts, USA
cariani@mac.com

1   INTRODUCTION

Arguably, the most important barrier to widespread use of automatic speech recognition systems in real–life situations is their present inability to separate speech of individual speakers from other sound sources: other speakers, acoustic clutter, background noise. We believe that critical examination of biological auditory systems with a focus on "reverse engineering" these systems, can lead to discovery of new functional principles of information representation and processing that can subsequently be applied to the design of artificial speech recognition systems.

From our experiences in attempting to understand the essentials of how the auditory system works as an information processing device, we believe that there are three major areas where speech recognizers could profit from incorporating processing strategies inspired by auditory systems. These areas are: 1) use of temporally–coded, front–end representations that are precise, robust, and transparent, and which encode the fine temporal (phase) structure of periodicities below 4 kHz, 2) use of early scene analysis mechanisms that form distinct auditory objects by means of common onset/offset/temporal contiguity and common harmonic structure (F0, voice pitch), and 3) use of central phonetic analyzers that are designed to operate on multiscale, temporally–coded, autocorrelation–like front–end representations as they present themselves after initial object formation/scene analysis processing. This paper will address the first two areas, with emphasis on possible neural mechanisms (neural timing nets) that could exploit phase–locked fine timing information to separate harmonic sounds on the basis of differences in their fundamental frequencies (harmonicity).

Psychoacoustical evidence suggests that the auditory system employs extremely effective low–level, bottom–up representational and scene analysis strategies to enable individual sound sources to perform this separation. Neurophysiological evidence suggests that the auditory system utilizes interspike interval information for representing sound in early stages of auditory processing. Interval–based temporal codes are known to provide high–quality, precise, and robust representations of stimulus periodicities and spectra over large dynamic ranges and in adverse sonic environments.

We have recently proposed neural timing networks that operate on temporally–coded inputs to carry out spike pattern analyses entirely in the time domain. These complement connectionist and time–delay architectures that produce "spatial", atemporal patterns of element activations as their outputs. In effect neural timing architectures provide neural network implementations of analog signal processing operations (e.g. cross–correlation, autocorrelation, convolution, cross–spectral product). The ubiquity of neural (tapped) delay lines in the brain may mean that many signal processing operations are more easily and flexibly implemented neurally using time domain rather than frequency domain and/or discrete feature detection strategies.

We have found that simple recurrent timing nets can be devised that operate on temporal fine structure of inputs to build up and separate periodic signals with different fundamental periods (Cariani, 2001a). Simple recurrent nets consist of arrays of coincidence detectors fed by common input lines and conduction delay loops of different recurrence times. A processing rule facilitates correlations between input and loop signals to amplify periodic patterns and segregate those with different periods, thereby allowing constituent waveforms to be recovered. The processing is akin to a dense array of adaptive–prediction comb filters. Based on time codes and temporal processing, timing nets constitute a new, general strategy for scene analysis in neural networks. The nets build up correlational invariances rather than using features to label, segregate and bind channels: they provide a possible means by which the fine temporal structure of voiced speech might be exploited for the speaker separation and enhancement.

2   PITCH AND AUDITORY SCENE ANALYSIS

Perhaps the most basic function of a perceptual system is to coherently organize the incoming flux of sensory information into separate stable objects (Bregman, 1981, 1990, Handel, 1989, Mellinger and Mont-Reynaud, 1996). In hearing, sound components are fused into unified objects, streams and voices that exhibit perceptual attributes, such as pitch, timbre, loudness, and location. Common periodicity, temporal proximity (onset, duration, offset), frequency, amplitude dynamics, phase coherence, and location in auditory space are some of the factors that contribute to fusions and separations of sounds.

For concurrent sounds, common harmonic structure plays perhaps the strongest role in forming unified objects and separating them (Mellinger and Mont-Reynaud, 1996, Hartmann, 1988). As a rule of thumb, voices and musical instruments having different fundamental frequencies (F0s) can be easily separated provided that their fundamentals differ by more than a semitone (6%, the F0 separation of two adjacent keys on a piano). Harmonic complexes with different fundamentals produce strong pitches at their fundamentals (even when the fundamental is not present in the power spectrum or when frequencies near the fundamental are masked out with noise). The mechanisms underlying pitch perception and auditory object formation therefore appear to intimately linked.

3   PROBLEMS WITH TRADITIONAL CHANNEL-BASED CODING OF THE AUDITORY STIMULUS

Traditionally, following the formidable intellectual synthesis of Helmholz and Fletcher, the auditory system has been conceived in terms of a central analysis of running frequency–domain representations of the stimulus power spectrum. Almost all extant front–end representations for automatic speech recognizers follow this assumption that a magnitude time–frequency spectrograph–like representation of the speech signal crudely mirrors the processing taking place in the central auditory system. Although is often conventionally assumed that the cochlea itself implements a array of narrowly–

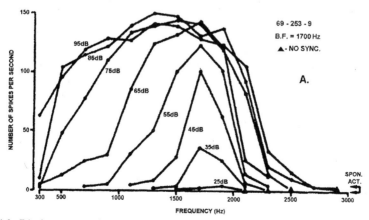

Figure 4.1. Discharge rate of a low spontaneous rate, high threshold auditory nerve fiber as a function of tone frequency and SPL. From Rose, 1971 (Rose *et al*, 1971).

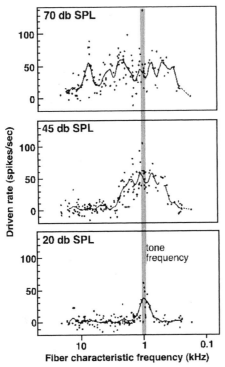

Figure 4.2. Driven rate–place profiles for a population of cat auditory nerve fibers of different characteristic frequencies in response to a 1 kHz tone presented at three levels. (Irvine, 1986) Data originally from Kim and Molnar (1979).

tuned bandpass filters that then subsequently create tonotopically–organized-frequency maps in the central auditory system, this picture of spectral coding presents many profound difficulties in light of what we know about neural response patterns at the level of the auditory nerve and the auditory CNS.

First, it has been known for decades now that the frequency response areas of auditory nerve fibers broaden dramatically as stimulus levels exceed 60 dB SPL (where tones below a fiber's characteristic frequency lie in the broad "tail" region of the fiber's tuning curve). A direct consequence of this is that simple representations of the stimulus spectrum based upon profiles of firing rates (whether absolute or driven) across whole neural populations perform best at low SPLs (<< 50 dB SPL). Such representations will be degraded at moderate to high SPLs as firing rates of fibers saturate. This is in marked contrast to auditory percepts, which almost invariably improve in precision and reliability at higher stimulus levels. In functional terms, this tonotopy is only viable as a scheme for encoding fine spectral distinctions at low sound levels. For frequencies more important for music and speech tonotopy is better conceptualized as "cochleotopy" – as a reflection of the most direct connections to the sensory receptor surface, rather than as a vehicle for frequency coding in its own right. Throughout the auditory neurophysiology literature, this general problem with the degradation of tonotopy, which is most acute for best frequencies below a few kHz, is seen at all levels of auditory processing from periphery to cortex. Two general (arguably ad-hoc) remedies have been proposed to save the rate–place picture: 1) the selective use of the fewer and fewer neural elements that exist that have higher and higher rate–thresholds and therefore dynamic ranges better suited to encode higher stimulus levels, and 2) reliance on lateral inhibition and the locations of shoulders of population–wide excitation patterns to infer the stimulus spectrum. Although these strategies can be made to work in particular cases (e.g. 2AFC discrimination), it is hard to envision how such representations would work in more general contexts to infer the spectra of novel stimuli and multiple sounds. It is even harder to imagine why pitch and timbre percepts based on such representations and analyses would be highly invariant with respect to stimulus intensity, with jnd's even improving at levels where units with appropriately high thresholds are relatively few.

Traditional approaches to auditory scene analysis also operate using the central spectrum assumption, that auditory objects are best described in terms of patterns of activation amongst frequency channels. In this view, the task of an auditory separation mechanism is channel segregation and binding, i.e. to label and segregate frequency channels according to pitch–related features (e.g. Meddis and Hewitt, 1992) and then to bind them together to form separate objects, streams, and voices.

4   TEMPORAL REPRESENTATION OF AUDITORY STIMULI

Historically, conceptions of the auditory system as a temporal pattern (periodicity) analyzer have developed alongside those that cast it as a frequency analyzer (Boring, 1942, Cariani, 1999, de Cheveigné, 2004). When one examines patterns of neural activity in the auditory nerve array (Fig. 4.4), one is immediately struck by the ubiquity of the temporal patterning of activity. In essence, the stimulus impresses its fine time structure on the temporal discharge patterns of multitudes of nerve fibers.

Periodicity–based theories of auditory representation account for the pitches of complex tones in terms of population–based all–order interspike interval statistics (Cariani and Delgutte, 1996, Cariani, 1999). The most common sets of intervals present in the whole population at any given time predict

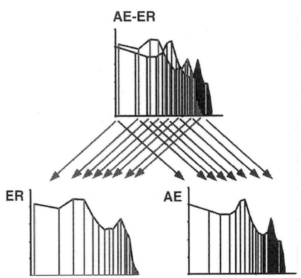

Figure 4.3. Traditional strategies for auditory scene analysis operate in the frequency domain to assign different sets of frequency channels to different auditory objects.

the pitch that will be heard. This account encompasses pitches evoked by pure and complex tones that have periodicities (frequencies, fundamentals) below the limits of strong phase–locking (~4 kHz). The one exception of which we are aware are click rate pitches produced by click trains of alternating polarity, where the click rate pitch is heard an octave above the fundamental. Periodicity–based representations effectively explain the frequency limits of musical tonality (octave matching, musical interval recognition), which only exists up to around 4 kHz (Burns, 1999).

For example, the familiar Big Ben melody can be played in different descending registers, beginning around 10 kHz. At about 4 kHz, the melody becomes recognizable. The level–invariant nature of low–frequency hearing (vs. the level–dependent character of high frequency hearing) strongly suggests that low– and high–frequency hearing rely on different neural mechanisms: a temporal mechanism for periodicities below 4 kHz, and an atonal, level–dependent mechanism for higher frequencies. Lower frequency spectra are also effectively represented in population–interval distributions. The reason is that each partial impresses its temporal structure on the temporal discharge patterns of fibers whose CFs are closest to it in frequency. The result is that the stimulus partitions cochlear territories according to the relative magnitudes of the various partials. This is illustrated in the ANF responses depicted in Figure 4.5, where dominant harmonics (associated with vowel formants) drive different CF territories. As a consequence, when all of

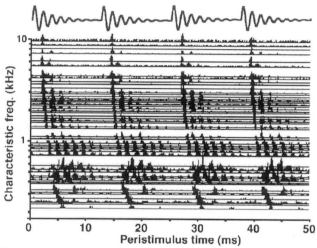

Figure 4.4. Temporal discharge patterns of auditory nerve fibers. Peristimulus time histogram (PSTH) responses of 52 cat auditory nerve fibers to a single-formant vowel (F0: 80 Hz, F1 = 640 Hz, 100x, 60 dB SPL). PSTH baselines indicate fiber characteristic frequencies.

the intervals are summed together into the population — interval distribution, the distribution resembles the autocorrelation function of the stimulus (for frequencies up to the limits of phaselocking). Place per se is therefore not strictly necessary to encode stimulus spectrum — each vowel with its characteristic formant structure and power spectrum produces a characteristic population– interval representation (that is precise and level–invariant). It is therefore possible and desirable to collapse correlogram–like representations across frequency in order to use the second dimension for time.

The result is an *autocorrelogram* (Fig. 4.6), which depicts the running autocorrelation of the stimulus. Although it is similar to a spectrogram in many respects, the autocorrelogram prominently depicts the dominant periodicities of speech sounds, i.e. voice pitch. We believe that effective front–end representations for speech analysis should represent those stimulus dimensions that are most prominent in speech perception, which we take to be voice pitch (dominant periodicity), vowel quality (spectrum), fast amplitude and frequency patterns (dynamic aspects of timbre), and the manner in which they are grouped together to form discrete voices and objects. In contrast with spectrograms, the autocorrelogram effectively depicts invariants associated with voice pitch, which is crucial for speaker separation and enhancement.

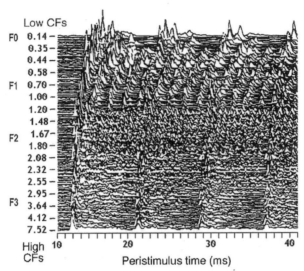

Figure 4.5. Responses of cat auditory nerve fibers to a multiformant vowel (/da/). Vertical axis: fiber CF (kHz), horizontal axis: peristimulus time (ms). Dominant harmonics in the respective formant regions impress their temporal structure on swaths of fibers, thereby partitioning cochlear (CF) territories. The production of intervals associated with the different partials thus reflects their relative magnitudes. From Seecker-Walker and Searle (1990), who analyzed data from Young and Sachs (1979).

5   PHASE–EFFECTS IN F0–BASED SEPARATION

It thus appears to us that the vast majority of the information that is utilized for perception of music and speech is temporal in nature. If we take this observation seriously, then new ways of thinking about the nature of auditory objects immediately present themselves. The ubiquity of phase–locking in early auditory processing means that spike timing patterns reflect the fine time structure (and phase structure) of the stimulus. This information is therefore potentially available for scene analysis mechanisms.

Although perception of pitch and timbre of stationary lower–frequency stimuli (the psychophysicists would say of resolved harmonics) is famously phase–insensitive, auditory grouping mechanisms are highly sensitive to abrupt phase changes. Harmonic complexes with F0s more than 2 semitones separate easily into two auditory objects with two distinct low (F0) pitches and timbres. We shall look at the double vowel case in more detail momentarily. Kubovy has demonstrated that abrupt changes in the phase and/or amplitude of a harmonic is sufficient to cause that harmonic to "pop–out". Each of these examples, including the separation of mistuned harmonics from complexes (see Kubovy, 1981, and Darwin and Gardner, 1986), suggests a separation mechanism based on a period-by-period comparison. A mechanism that compared the waveform (spike pattern) of a preceding period with that of a subsequent period would register an ongoing disparity, and the form of this disparity would be that of the pure tone component. Assuming that pure tones are encoded temporally, then the auditory CNS would interpret this spike pattern disparity as a pure tone.

Double vowels with same and different fundamental periods would presumably be analyzed by this mechanism. The psychophysics of their separation and identification has been studied extensively (Summerfield and Assmann, 1991, Assmann and Summerfield, 1990). Models had been proposed (Meddis and Hewitt, 1992, de Chevigné, 1999), and their neural responses had been investigated (Cariani and Delgutte, 1993, Cariani, 1995, Palmer, 1988, 1992). When the two vowels have the same F0, they fuse into one object, when the two F0s differ by a semitone or more, the vowels separate. When the F0s are the same, listeners are able to correctly identify both individual vowel constituents about half the time. For F0 separations of more than a semitone, listeners improve by 15–20%. A mechanism that built up recurrent temporal patterns would behave in a qualitatively similar way,

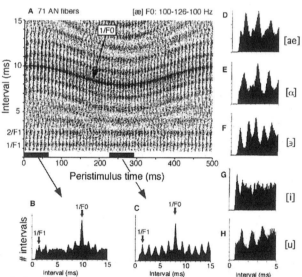

Figure 4.6. Population–interval representations of five synthetic vowels compiled from responses of 50-100 cat auditory nerve fibers. A. Population autocorrelogram for a variable-F0 vowel /ae/, F0= 100-126 Hz. The dark interval band closely follows the fundamental period. B–C. Cross sections of the autocorrelogram are population–interval distributions for 60 ms stimulus segments. The largest interval peak corresponds to voice pitch, while patterns of short intervals (< 5 ms) reflect dominant harmonics (formant structure). D–H. Population–interval distributions for five synthetic vowels, averaged over F0s 100-126 Hz.

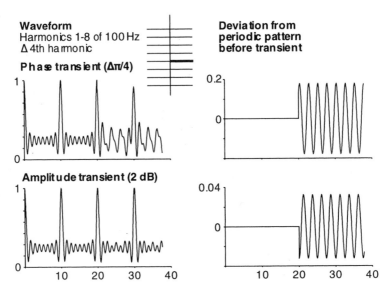

Figure 4.7. Phase and amplitude transients that cause individual harmonics to temporarily "pop out" of harmonic complexes. Left: waveforms of harmonic complexes having abrupt phase and amplitude shifts at 20 ms. Right: Subtraction of last two periods of the waveforms from the first two (period-by-period deviation from perfectly periodic waveforms).

fusing the two patterns when they have the same fundamental period and separating them when they have different ones.

A visual analogy can be constructed by overlaying two transparencies with arbitrary abstract patterns on them (e.g. random dots). When the transparencies are moved together, the patterns fuse, when they are moved independently, their respective patterns immediately separate. One intuits that there is a mechanism that builds up the respective invariant patterns of dots in the transparencies and integrates these images when they are translated spacially. The mechanism will treat the two transparencies as one when the relations of dots on each sheet is stable relative to the dots on the other. Likewise, it will find two sets of invariant relations when the sheets are moved relative to each other — while relations between dots on the two transparencies are rendered unstable and varying, those within a given sheet remain stable. In the auditory case, the relations are between patterns of spikes associated with the two vowels. The conventional visual scene analysis explanation assumes representations of the patterns as sets of active feature detectors (e.g. motion detectors, bug detectors) the scene analysis task being to separate the features associated with the two dot patterns on the transparencies. In the double vowel example, we know from neurophysiological experiments and computer simulations that two vowels separated by a few

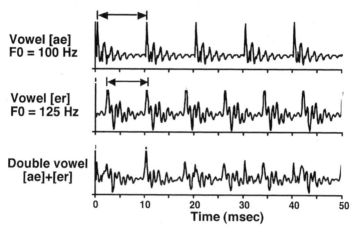

Figure 4.8. Double vowel stimuli.

semitones in frequency drive the same neuronal frequency channels in the auditory nerve. The spike timing information that encodes the two vowels is temporally multiplexed in the same auditory nerve fibers (Figure 4.9). This suggests that the auditory system must use phase/fine timing information to separate the two vowels. Back to the visual case, since visual neurons also "phase–lock" to moving images, it is not out of the realm of possibility that the visual system utilizes spatial patterns of temporally–correlated spikes to represent and separate forms in an analogous manner (Cariani, 2004, in press). This temporal correlation hypothesis potentially explains why motion is essential for vision: why visual forms disappear when their positions are stabilized on the retina.

One possibility then is that auditory objects are first formed using a low–level phase–sensitive mechanism that builds up periodic patterns and separates divergent ones. Such a mechanism would operate on the temporal coherence of fine time patterns (internal phase coherence) rather than detecting discrete features and sorting channels on that basis. Once objects were formed by such a mechanism, they would subsequently be analyzed by an phase–insensitive mechanism (pitch, timbre, loudness, location). Recurrent neural timing nets are an attempt to demonstrate how this strategy for scene analysis could be implemented neurally.

We strongly believe that new information–processing models that focus on functional principles are absolutely essential for long–term progress in auditory neuroscience and speech recognition. In auditory neuroscience, it seems likely that computational biophysical simulations aimed at accounting for the input–output behavior of particular neuronal elements will not lead us

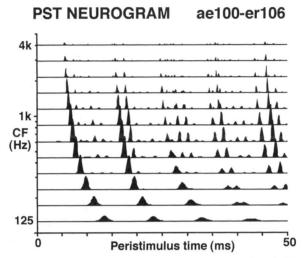

Figure 4.9. Simulated auditory nerve response to a double vowel with F0s separated by a semitone (6%). Information associated with the respective vowels is multiplexed in the time domain. The situation is similar for larger F0 separations.

to the functional principles we need to understand how the auditory system works as an information–processing system. In lieu of strong functional hypotheses, the neurophysiologists will be left to creating ever larger catalogs of neural responses. Audio separation and speech recognition should pay attention to natural auditory systems, if only because these are still by far the most effective sound processing and speech recognition devices on the planet. Again the attention needs to focus on essential functional principles – which aspects of auditory function produce the performances that we see.

To date, there have been only a few isolated attempts to exploit the power of temporal codes in providing high–quality general–purpose auditory front–ends for speech recognizers (Ghitza, 1992, 1988). In the last decade correlo-grams have been used effectively to label and separate subsets of frequency channels (Wang and Brown, 1999). While these efforts, in incorporating fine timing information, represent a great improvement over purely spectral analy-ses, it seems to us unlikely that the auditory system operates in this way to bind together discrete frequency channels. When we contemplate the image of spike activity in a simulated auditory nerve (Fig. 4.9) that uses broader, more physiologically realistic cochelar filters rather than much narrower "auditory filters" derived from (whole system) psychophysics, it seems immediately obvious that there are no discrete, narrowly-tuned neural frequency channels to be sorted. The situation is no better at higher levels of processing. Yet to be produced is neurophysical evidence suggesting that two low-frequency pure tones 20% apart in frequency (easily heard as two discrete tones) will excite separable populations of neurons in higher auditory stations. The means by which the central auditory system achieves fine low-frequency discrimination and selectivity remains a completely open (and largely neglected) question for auditory neuroscience. It is difficult to envision how investigations of scene analysis mechanisms at the cortical level can proceed without some prior understanding of the precise nature of neural representations of periodicity and spectrum at that level. On the other hand, well-constructed studies of neural scene analysis mechanisms (e.g. Sinex, 2002) could conceivably shed precious light on the nature of these representations at higher centers.

In our view, temporal coding is not simply a special-purpose "hack" to do F0-based scene analysis, but a fundamental organizing principle of vertebrate auditory systems. Information processing in the auditory system may well be based on fine timing information that our present spectrographic representations throw away at an early stage. If neural analysis of sound primarily involves time-domain operations then it behooves us to explore similar strategies for sound analysis in artificial systems. This line of thinking suggests wider and more general use of correlograms and autocorrelograms for sound separation, front-end representations, and even back-end recognition strategies. Ideally we should use such representations in conjunction with processing strategies that parallel human auditory scene analysis. Here the most important mechanisms group sounds by temporal contiguity (common onset/offset) and by common harmonicity (F0). When we have both auditory representations and sound separation strategies that exploit the temporal microstructure of sounds, then we should be well on our way to developing much more robust bottom-up automatic speech recognition systems.

6 RECURRENT TIMING NETS

For these and other reasons, we have strived to develop new heuristics for how auditory images might be formed and separated. Both feedforward and recurrent networks we have been considered and their basic computational properties were explored (Cariani, 2001a, 2001b). Neural timing nets demonstrate how analog time–domain filtering operations could conceivably be performed in neural network implementations. They expand the realm of possible signal–processing mechanisms available to nervous systems. We hope they will have the effect of catalyzing new functional hypotheses for how information could be represented, transmitted, multiplexed, broadcast, analyzed, and integrated. These networks have also been investigated in the context of music perception (Cariani, 2002), as possible approaches to tonality and rhythmic induction.

Recurrent timing networks were inspired in different ways by models of stabilized auditory images (Patterson *et al.*, 1995), neural loop models (Thatcher and John, 1977), adaptive timing nets (MacKay, 1962), adaptive resonance circuits (Grossberg, 1988), the precision of echoic memory, and the psychology of temporal expectation (Jones, 1976, Miller and Barnet, 1993). Although much is known about time courses of temporal integration that are related to auditory percepts (pitch, timbre, loudness, location, object separation, and various masking effects), we currently have few good models for how incoming information in the auditory periphery is integrated over time by the central auditory system to form stabilized auditory percepts. If the information involved is indeed temporally–coded, then architectures that store temporal patterns in reverberating circuits eventually come to mind. One envisions the signals themselves circulating in closed transmission loops or regenerated via cellular recovery mechanisms. These temporal memory traces (temporal echoic memories) would be compared with incoming patterns via coincidence–detectors that compute temporal correlations. Neural representations would thus build up over time, dynamically creating sets of perceptual expectations that could either be confirmed or violated. Periodic signals, such as rhythms, would thereby create strong temporal expectancies (Cariani, 2002, Fraisse, 1978).

The simplest recurrent timing networks imaginable consist of a 1-D array of coincidence detectors having common direct inputs (Figure 4.10). The output of each coincidence element is fed into a recurrent delay line such that the output of the element at time t circulates through the line and arrives tau milliseconds later (the signal that arrives back at time t is the one that was emitted

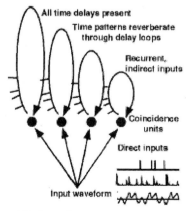

Figure 4.10. Simple recurrent timing net.

at t–tau). A processing rule governs the interaction of direct and circulating inputs.

In their development the networks have evolved from simple to more complex. In the first simulations (Cariani, 2001b), binary pulse trains (resembling spike trains) with repeated, randomly selected pulse patterns (e.g. 100101011-100101011-100101011...) were passed through the network. For each time step, incoming binary pulses were multiplied by variable–amplitude pulses arriving through the delay loop. In the absence of a coincidence with a circulating pulse, the input pulse was fed into the delay loop without facilitation. When coincidences between incoming and circulating pulses occurred, the amplitude of the circulating pulse was increased by 5% and the pulse was fed back into the loop. It was quickly realized that such networks rapidly build up any periodic pulse patterns in their inputs, even if these patterns are embedded amidst many other pulses. A periodic pattern invariably builds up in the delay loop whose recurrence time matches its repetition time. Thus, recurrent time patterns are repeatedly correlated with themselves to build up to detection thresholds. In effect, these autocorrelating loops dynamically create matched filters from repeating temporal patterns in the stimulus. In this manner, temporal–pattern invariances are enhanced relative to uncorrelated patterns – the network functions as a pattern amplifier. When two repeating temporal patterns each with its own repetition period were summed together and presented to such nets, the two patterns emerged in the two different delay loops that had recurrence times that corresponded to the repetition periods of the patterns. Although the proportional facilitation rule distorted signal amplitudes, the temporal patterns of pulses corresponding to the two rhythms could be recovered in the circulating waveforms. A neural network can therefore carry out an analog–style separation of signals in the time–domain. To do this, inputs need to be temporally coded, processing elements must have sufficiently narrow coincidence windows, delays must be relatively precise, and processing rules must be judiciously chosen.

7   SEPARATION OF DOUBLE VOWELS

Although binary pulse trains resemble spike trains of individual neurons, most real neural information processing appears to be carried out by large ensembles of neurons working in concert. Subsequent simulations (Cariani, 2001b) therefore used positive real–valued input signals that qualitatively resemble neural post–stimulus time histograms (e.g. time series of spike counts that would be produced by an ensemble of similar neural elements whose discharges were stimulus–locked). Proportional facilitation was replaced by a processing rule that adaptively adjusted the output signal in a more graceful and less distorting manner. Multichannel implementations subsequently processed the double vowels using an auditory nerve front end with 24 CF channels. The instantaneous spike rate of each frequency channel was fed into an array of delay loops and the autocorrelations of the circulating waveforms in corresponding delay loops were combined to theroduce simulated population–interval distributions.

Our current single–channel implementation uses a simple error–adjustment processing rule that can operate on signals with both positive and negative values. $(H(t) = H(t - \text{tau}) + B_{\text{tau}}[X(t) - H(t - \text{tau})])$ describes the input-output function of each processing element. B_{tau} determines the rate of adjustment, and its dependence on $B_{\text{tau}} = \text{tau}/33\,\text{ms}$ ensures that shorter loops are not favored.

To a signal processing engineer, the net somewhat resembles a temporally-coded neural implementation of a bank of comb filters, albeit ones with very short (1-2 period) temporal integration times. It would be a mistake to dismiss them as simple autocorrelations (in the same way that it would be a mistake to reduce all the refinements of spectral analysis to Fourier's Theorem). All of the most effective strategies for implementing F0-based separations (comb filters, correlograms, cancellation operations) are in one way or another formally related to autocorrelation. But with their short memories, time-domain implementations, and avoidance of early windowing, these nets have more in common with the sample-by-sample and period-by-period harmonic cancellation strategies of de Cheveigne (Chapter 16, this volume) than they do with traditional sharp comb filters that utilize long integration times.

Synthetic, three–formant double vowels (/ae/, /er/) with different fundamentals (100, 112 Hz) were summed and processed by the network (Fig. 4.11). The signals circulating in the 150 delay loops are shown in the response map, where it can be seen that the recurrence times of the loops with the highest average signal strength correspond to the periods of the two vowels (8.9 and 10 ms). The signals circulating in these two delay channels after 70 ms of processing highly resemble the two vowel constituents. Correlations between the autocorrelations of these processed signals and those of the individual-vowels show how the signal separation unfolds over processing time.

In both single– and multi–channel cases, when vowel fundamentals were separated by a semitone or more, the autocorrelations (and hence, power spectra) of the constituent vowels could be accurately recovered. The quality of the separations improved as a function of ΔF0 and vowel duration. The multichannel simulation demonstrated how recurrent timing nets could be scaled

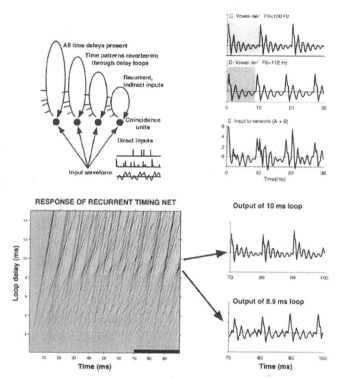

Figure 4.11. Separation of a double vowel (ae-er) with different F0s (100, 112 Hz) into its constituent waveforms by means of a recurrent timing net. Top. Array of processing elements and delay loops. Individual synthetic vowels and their summed waveforms. Bottom. Response of the timing network to the double vowel. Arrows emerge from the two loops with the biggest signals. Waveforms circulating in the two loop channels after 70 ms.

up to process multichannel positive, real–valued signals not qualitatively unlike those produced by auditory nerve arrays. It also showed that auditory objects can be separated when they activate the same sets of broadly tuned frequency channels (i.e. without usable rate–place information), provided that phase–locked fine timing information is available. The information related to multiple auditory objects (the two vowels) is embedded in the phase structure of the stimulus and phase–locked neural responses. This is a relatively straightforward auditory example of how information can be multiplexed in the time domain. The networks also demonstrate how an auditory scene analysis system could exploit phase–coherence and F0–differences without first carrying out explicit estimations of F0 and segregating frequency channels on that basis. For example, in Wang and Brown (1999), correlograms (f, tau) label frequency channels with common F0–related autocorrelation profiles, which are then grouped using an array of synchronizing oscillators. In timing

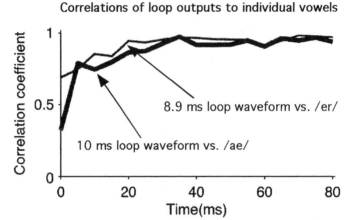

Figure 4.12. Performance of the network in separating the vowels. Degree of similarity between the output waveforms and the individual vowel constituents, as measured by the correlation between their autocorrelations. The network rapidly separates the two vowels, within 2-3 periods.

networks, formation of auditory objects occurs prior to analyses that yield object attributes (F0 pitch, timbre).

One can ask how well these networks handle more than two auditory objects. A third vowel /ee/ with yet a different fundamental (125 Hz) was added to the mixture. This is akin to the problem of hearing out three different kinds of musical instruments playing different notes (first that there are three different notes, second that instruments with different timbres are playing the three notes). Processing by the network resulted in the appearance of another strong signal in the response map. Separation of the signals in the three–vowel case was somewhat slower than for two vowels, but there was only a slight reduction in the final quality of the separated signals. The performance of the network therefore appears to be highly robust.

We are developing RTN-based voice pitch trackers that can handle running speech (Fig. 4.13). The general form of the RTN response looks somewhat similar to the stimulus autocorrelogram, and the RTN has no difficulty following the rapidly varying pitch contour of the speaker. F0 tracks can be constructed over multiple time scales (samples, loop periods, syllables, sentences) and the signals in the corresponding loop-time trajectories can be easily and relatively seamlessly assembled on a sample-by-sample basis without resorting to windowing and without leaving the time domain. Athough F0-tracking only separates voiced segments, voicing provides a temporal framework for delimiting unvoiced segments. If voiced segments of different speakers have asynchronous onsets and offsets, it may be possible to use continuity rules to assign unvoiced segments to different speakers. Once voiced segments can be identified, then adjacent voiced-unvoiced or unvoiced-voiced patterns may also be analyzable as well-formed CV units.

8   ENHANCEMENT OF VOWELS IN NOISE

A possible use of recurrent timing nets is for enhancement of periodic sounds in noisy environments. Such processing would be useful for processing music and voiced speech. Reductions in effective S/N ratios could be expected to improve speech reception by human listeners and automatic recognition by machines. Related kinds of correlation–based strategies were used in the 1950's to detect periodic signals in noise (Lange, 1967, Meyer-Eppler, 1953), in situations where the period of the target signal was known a priori. The present networks systematically sample all possible delays, such that the optimum delay(s) can be determined by choosing the loop(s) with the largest signal rms.

In order to assess the performance of the network in noise, a synthetic, three–formant vowel (/ae/, F0=100 Hz, S/N = -20-20 dB) was added to frozen white noise at different S/N ratios that ranged from –20 to 20 dB. The input and output signals from the optimum delay loop (tau = 10 ms) are shown in the top panels of Figure 4.8. The bottom panels show the correlation between the autocorrelation of these signals and that of the vowel in near–quiet (20 dB S/N). Similarities between the processed signals and the minimal–noise case improve with S/N and processing time. For all S/N ratios less than 1, the network produced output signals (thick curves) that had higher correlations than the unprocessed, input signals (thin curves). Processing by the network shifts the curves to the left, an improvement in S/N by roughly 4-10 dB that is comparable to improvements that have been reported using comb filters (Stern, 2003).

9   THEORETICAL CONSIDERATIONS

We have been contemplating some of the longer–range theoretical implications that relate to brains as general–purpose self–organizing correlation machines that extract invariant patterns in their inputs. Recurrent timing nets use the periodic patterns in their inputs to dynamically form matched templates that they compare with subsequent inputs. For ease of visualizing their

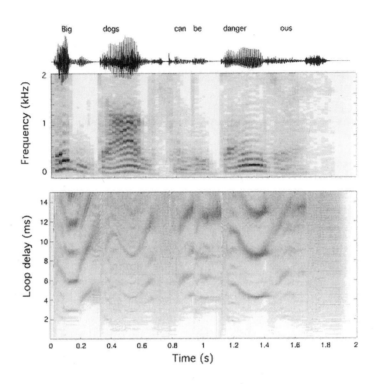

Figure 4.13. Response of a recurrent timing net (RTN) to running speech. Top: waveforem and spectrogram for HINT sentence, "Big dogs can be dangerous". Bottom: Output of RTN after smoothing and normalization. Contours related to the voice pitch are seen in loop channels near 9 ms.

behavior, we have considered ordered arrays of monosynaptic delay loops. It is conceivable, however, that such processing can also be carried out in randomly connected networks, provided that recurrent, multisynaptic pathways are available that span a wide range of loop–delays. If such networks use coincidence elements that are transiently facilitated by temporal coincidences, then it is not hard to envision how they might support dynamically–formed reverberatory memories capable of retaining temporal patterns and interspike interval statistics. In their operation such networks would be akin to self–organizing recurrent synfire chains (Abeles, 1990, 2004) in which both synchrony and temporal patterning of spikes play critical roles. Processing using spike statistics "liberates signals from the wires", since the respective identities of signals then no longer depends on which particular input lines the respective signals conveyed. It opens the possibility of more flexible kinds of neural networks that can multiplex signals, broadcast them in novel ways.

= Dave's 1-channel "beanforming"

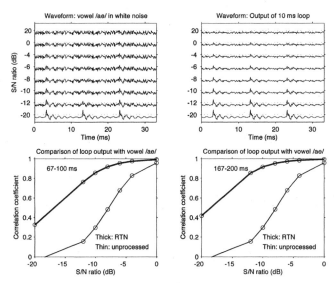

Figure 4.14. RTN representation of a vowel in white noise.

10   ACKNOWLEDGMENTS

We thank Pierre Divenyi for his encouragement. This work was supported in part by the National Science Foundation, NSF-EIA-BITS-013807.

References

Abeles, M., 2004. *Time is precious*. Science **304** (**23 April**): 523-524.

Abeles, M., 1990. *Corticonics*. Cambridge University Press, Cambridge.

Assmann, P.F. and Summerfield, Q., 1990. Modeling the perception of concurrent vowels: Vowels with different fundamental frequencies. *J. Acoust. Soc. Am.*, **88**: 680–697.

Boring, E.G., 1942. *Sensation and Perception in the History of Experimental Psychology*. Appleton-Century-Crofts, New York.

Bregman, A.S., 1981. *Asking the "what for" question in auditory perception. in, Perceptual Organization*. M. Kubovy and J.R. Pomerantz (Eds). Lawrence Erlbaum Assoc., Hillsdale, NJ, pp. 99–118.

Bregman, A.S., 1990. *Auditory Scene Analysis: The Perceptual Organization of Sound*. MIT Press, Cambridge, MA, 773 pp.

Cariani, P., 1995. As if time really mattered: temporal strategies for neural coding of sensory information. *Communication and Cognition – Artificial Intelligence (CC–AI)*, **12(1–2)**: 161–229 (Reprinted in: K Pribram, ed. Origins: Brain and Self–Organization, Hillsdale, NJ: Lawrence Erlbaum, 1994, 208–252.).

Cariani, P., 1999. Temporal coding of periodicity pitch in the auditory system: an overview. *Neural Plast*, **6(4)**: 147-72.

Cariani, P., 2001a. Neural timing nets. *Neural Networks*, **14(6–7)**: 737–753.

Cariani, P., 2001b. Neural timing nets for auditory computation, in *Computational Models of Auditory Function*. S. Greenberg and M. Slaney (Eds). IOS Press, Amsterdam, pp. 235–249.

Cariani, P., 2002. Temporal codes, timing nets, and music perception. *J. New Music Res.*, **30(2)**: 107–136.

X Cariani, P., 2004 (in press). Temporal codes and computations for sensory representation and scene analysis. *IEEE Trans.on Neural Networks, Special Issue on Temporal Coding for Neural Information Proc.*

Cariani, P. and Delgutte, B., 1993. Interspike interval distributions of auditory nerve fibers in response to concurrent vowels with same and different fundamental frequencies. *Assoc. Res. Otolaryngology. Abs.*: **373**.

Darwin, C.J. and Gardner, R.B., 1986. Mistuning a harmonic of a vowel: grouping and phase effects on vowel quality. *J Acoust Soc Am*, **79(3)**: 838–45.

de Chevigné, A., 1999. Waveform interactions and the segregation of concurrent vowels. *J Acoust Soc Am*, **106(5)**: 2959–72.

de Cheveigné, A., 2004. The cancellation principle in acoustic scene analysis. *This volume.*

de Cheveigné, A., 2004, in press. Pitch perception models. in, *Pitch*, C.J. Plack and A.J. Oxenham (Eds). Springer Verlag, New York.

Fraisse, P., 1978. Time and rhythm perception. in, *Handbook of Perception. Volume VIII. Perceptual Coding*, E.C. Carterette and M.P. Friedman (Eds). Academic Press, New York, pp. 203–254.

Ghitza, O., 1988. Temporal non–place information in the auditory–nerve firing patterns as a front–end for speech recognition in a noisy environment. *J. Phonetics*, **16**: 109–123.

Ghitza, O., 1992. Auditory nerve representation as a basis for speech processing. in, *Advances in Speech Signal Processing*, S. Furui and M.M. Sondhi (Eds). Marcel Dekker, New York, pp. 453–485.

Grossberg, S., 1988. *The Adaptive Brain, Vols I. and II.* Elsevier, New York.

Handel, S., 1989. *Listening: An Introduction to the Perception of Auditory Events.* MIT Press, Cambridge, 597 pp.

Hartmann, W.M., 1988. Pitch perception and the segregation and Integration of auditory entities. in, *Auditory Function: Neurobiological Bases of Hearing*, G.M. Edelman (Ed) John Wiley & Sons, New York, pp. 623–347.

Irvine, D.R.F., 1986. The Auditory Brainstem. *Progress in Sensory Physiology 7.* Springer–Verlag, Berlin, 279 pp.

Jones, M.R., 1976. Time, our lost dimension: toward a new theory of perception, attention, and memory. *Psychological Review*, **83(5)**: 323–255.

Kim, D.O. and Molnar, C.E., 1979. A population study of cochlear nerve fibers: comparison of spatial distributions of average–rate and phase–locking measures of responses to single tones. *J. Neurophysiol.*, **42(1)**: 16–30.

Kubovy, M., 1981. Concurrent–pitch segregation and the theory of indispensable attributes. in, *Perceptual Organization*, M. Kubovy and J.R. Pomerantz (Eds). Lawrence Erlbaum Assoc., Hillsdale, NJ, pp. 55–98.

Lange, F.H., 1967. *Correlation Techniques*. Van Nostrand, Princeton, 464 pp.

MacKay, D.M., 1962. Self–organization in the time domain. in, *Self–Organizing Systems 1962*, M.C. Yovitts, G.T. Jacobi and G.D. Goldstein (Eds). Spartan Books, Washington, D.C., pp. 37–48.

Meddis, R. and Hewitt, M.J., 1992. Modeling the perception of concurrent vowels with different fundamental frequencies. *J. Acoust. Soc. Am.*, **91**: 233–245.

Mellinger, D.K. and Mont-Reynaud, B.M., 1996. Scene analysis. in, *Auditory Computation*, H. Hawkins, T. McMullin, A.N. Popper and R.R. Fay (Eds). Springer Verlag, New York, pp. 271–331.

Meyer-Eppler, W., 1953. Exhaustion methods of selecting signals from noisy backgrounds. in, *Communication Theory*, W. Jackson (Ed) Butterworths, London, pp. 183–194.

Miller, R.R. and Barnet, R.C., 1993. The role of time in elementary associations. *Current Directions in Psychological Science*, **2(4)**: 106–111.

Palmer, A.R., 1988. The representation of concurrent vowels in the temporal discharge patterns of auditory nerve fibers. in, *Basic Issues in Hearing*, H. Duifhuis, J.W. Horst and H.P. Wit (Eds). Academic Press, London, pp. 244–251.

Palmer, A.R., 1992. Segregation of the responses to paired vowels in the auditory nerve of the guinea pig using autocorrelation. in, *The Auditory Processing of Speech*, S.M.E.H. (Ed) Mouton de Gruyter, Berlin, pp. 115–124.

Patterson, R.D., Allerhand, M.H. and Giguere, C., 1995. Time–domain modeling of peripheral auditory processing: A modular architecture and a software platform. *J. Acoust. Soc. Am.*, **98(4)**: 1890–1894.

Rose, J.E., Hind, J.E., Brugge, J.R. and Anderson, D.J., 1971. Some effects of stimulus intensity on response of single auditory nerve fibers of the squirrel monkey. *J Neurophysiology*, **34(4)**: 685–699.

Secker-Walker, H.E. and Searle, C.L., 1990. Time–domain analysis of auditory–nerve–fiber firing rates. *J. Acoust. Soc. Am.*, **88(3)**: 1427–1436.

Sinex, D.G., Sabes, S.J. and Li, H., 2002. Responses of inferior colliculus neurons to harmonic and mistuned complex tones. *Hear. Res.*, **168**: 150-62.

Stern, R. 2003. Signal separation motivated by auditory perception. *Perspectives on Speech Separation*, Montreal, October 30- November 1, 2003. http://www.ebire.org/speechseparation.

Summerfield, Q. and Assmann, P.F., 1991. Perception of concurrent vowels: effects of harmonic misalignment and pitch–period asynchrony. *J. Acoust. Soc. Am.*, **89(3)**: 1364–1377.

Thatcher, R.W. and John, E.R., 1977. *Functional Neuroscience, Vol. I. Foundations of Cognitive Processes*. Lawrence Erlbaum, Hillsdale, NJ, 382 pp.

Wang, D.L. and Brown, G.J., 1999. Separation of speech from interfering sounds based on oscillatory correlation. *IEEE Trans. Neural Networks*, **10(3)**: 684–697.

Young, E.D. and Sachs, M.B., 1979. Representation of steady–state vowels in the temporal aspects of the discharge patterns of populations of auditory nerve fibers. *J. Acoust. Soc. Am.*, **66(5)**: 1381–1403.

Chapter 5

Blind Source Separation Using Graphical Models

Te-Won Lee
Institute for Neural Computation
University of California, San Diego
tewon@ucsd.edu

1   INTRODUCTION

Independent Component Analysis (ICA) (Attias, 1999, Bell and Sejnowski, 1995, Comon, 1994, Hyvärinen, Karhunen and Oja, 2002, Lee, 1998) is now a well established method for data analysis. Its popularity is due to its simple model with a wide range of interesting theories and its applicability to many real data analysis problems. Research directions in ICA are twofold and aimed at the relaxation of strong assumptions in traditional ICA methods (no sensor noise, square mixing matrix, known number of sources, independence of sources) as well as the use of ICA methods for low level signal representation. For the former, we propose the ICA mixture model and the variational Bayesian ICA model (Chan *et al.*, 2002) as a nonlinear extension of ICA that can handle sensor noise, estimate the number of sources, and model dependencies in the data. For the latter, ICA has been used as a tool for efficiently encoding speech signals for subsequent pattern recognition, compression and other machine learning tasks. In applying the algorithm to find a representation for speech signals, the learned speech basis functions are used for encoding speech signals for speech and speaker recognition tasks as well as the difficult problem of separating mixed sounds given only one channel. In this summary, we present these two directions in a graphical model by providing examples of source separation for one channel (simulations) and two channels (real recordings).

2   GRAPHICAL MODEL FRAMEWORK

Recently, new algorithms have been proposed to solve difficult signal processing problems. In many cases these algorithms can be described in a graphical model, which provides a general framework that allows the extension of algorithms to model new variables, parameters, signals and relationships amongst each other (Jordan *et al.*, 1998). The learning rules and algorithms for the new model can be developed in a principled mathematical manner using tools from statistical learning theory, graphical models and signal processing. The marriage between graphical models and signal processing methods is gaining acceptance in several research disciplines and successful examples include Hidden Markov Models (HMM) for speech recognition, Independent Component Analysis (ICA) for blind signal separation, error correction codes or turbo codes in communication systems, and probabilistic algorithms in robots.

2.1   Source Models

A single audio signal can be modeled with its probabilistic representation, the time varying structure, and its decomposition into fundamental basis functions that produce an efficient coding scheme. The generative model for a single source can be extended into a multiple source observation problem. The problem is then to understand the relationship between sources and how to model their interaction with little a priori knowledge. Blind source separation is a prime example for modeling multiple sources in an environment. Furthermore, the model is realistic since audio signals do not occur isolated but are active simultaneously. Multiple source models may be given for single channel observations as well as multiple channel observations. To model the interactions with changing environments, this multiple source model needs to be further extended to include contextual changes due to the environment or non-stationary character of the sources. The model should be able to make inference about the environmental dynamics, possibly track signal sources, and understand the structure of the interacting source signals.

In its simplest form, a source can be a random variable with a fixed probability density function. A non-linear function such as the sigmoid function or the tanh–function could represent the cumulative density for the source signal. In Bell and Sejnowski (1995) this non-linear function was used to separate super-Gaussian sources. This was a sufficient model because the goal was to estimate an unmixing matrix and the observation model was linear, deterministic (no sensor noise) and there were as many sources as given observation channels. There are many ways to extend this source model to include other density functions such as sub-Gaussian sources (Lee, 1998) and more complicated source densities that can be modeled with a mixture of Gaussians (MoG) (Attias, 1999).

Natural signals however are not random. To the contrary, they can have simple as well as complicated time structure. Speech signals for example are time–varying signals with correlations in time. One popular way is to model

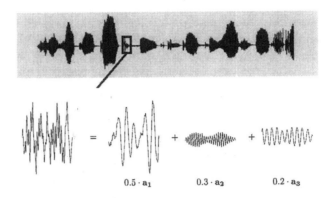

$$0.5 \cdot a_1 \qquad 0.3 \cdot a_2 \qquad 0.2 \cdot a_3$$

Figure 5.1. A speech segment is linearly decomposed into basis functions a_i and the corresponding coefficients.

these dependencies in an HMM. The parameters of the HMM are trained on speech data and different sets of parameters can provide models for phonemes. A different approach to modeling the time structure of the source is to learn the basis functions for the signal (Bell and Sejnowski, 1996). This is a data generative model for the speech signal in which the observed speech segment can be decomposed into learned basis functions and their corresponding coefficients. The basis functions are adapted such that coefficients are statistically independent, resulting in an efficient code.

In the subsequent sections, we illustrate how this data generative principle can be used to in blind separation problems in the case of one and two microphones.

3   ONE–CHANNEL SOURCE SEPARATION

The main concept behind the blind signal separation when given only a single channel recording is based on exploiting *a priori* sets of time–domain basis functions learned by ICA to the separation of mixed source signals observed in a single channel (Jang and Lee, 2003). The inherent time structure of sound sources is reflected in the ICA basis functions, which encode the sources in a statistically efficient manner. We derive a learning algorithm using a maximum likelihood approach given the observed single channel data and sets of basis functions. For each time point we infer the source parameters and their contribution factors. This inference is possible due to prior knowledge of the basis functions and the associated coefficient densities. A flexible model for density estimation allows accurate modeling of the observation and our experimental results exhibit a high level of separation performance for simulated mixtures.

The single channel blind source separation can be formulated as follows:

$$y = \lambda_1 x_1 + \ldots + \lambda_i x_i + \ldots + \lambda_N x_N \tag{1}$$

This kind of model as been studied extensively in the computational auditory scene analysis (CASA) literature (Brown and Cooke, 1994). The dominant approach includes the use of strong prior information about frequency clustering and the robust extraction of speech features. In the synthesis model, the observed signal y is generated by independent source signals x with different factor loadings λ. The goal is to infer the unknown source signals. This problem is highly ill conditioned and solutions can be formulated only for a constrained setting. The main idea behind our generative model approach is to make use of prior information as provided by the statistical structure of the signals of interest. The constraint is to obtain an overall efficient coding scheme, where the source signal prior information is obtained by a sparse decomposition of the signal through basis functions that have been learned. In the process of inferring the decomposition of the mixed signals, the parameters that model the linear generation of the independent source as well as the linear mixing of two sources given their basis functions and corresponding pdf structure are estimated via gradient ascent on the maximum a posteriori cost function. Figure 5.2 shows an example for separating two sources from a single observation. The details of the applied methods are described in Jang and Lee (2003).

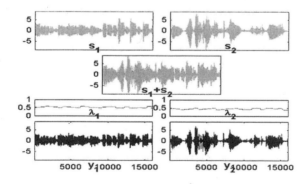

Figure 5.2. Single channel blind source separation. Separation results of jazz music and male speech. In vertical order: original sources (x1 and x2), mixed signal (x1+ x2), and the recovered signals.

3.1   Discussion on Single Channel

The technique in Jang and Lee (2003) for single channel source separation utilizes the time–domain learned ICA basis functions. Instead of traditional prior knowledge of the sources, the statistical structures of the sources that are inherently captured by the basis and its coefficients from a training set are exploited. The algorithm recovers original sound streams through gradient–ascent adaptation steps pursuing the maximum likelihood estimate, computed by the parameters of the basis filters and the generalized Gaussian distributions of the filter coefficients. With the separation results of the real recordings as well as simulated mixtures, the proposed method is applicable to real world problems such as blind source separation, denoising, and restoration of corrupted or lost data. We are interested in including the extension of this framework to perform model comparisons to estimate the optimal set of basis functions to use given a dictionary of basis functions. This is achieved by applying a variational Bayes method (Chan *et al.*, 2002) to compare different basis function models to select the most likely source. This method also allows us to cope with other unknown parameters such the as the number of sources. Other approaches to single channel source separation can be found in deChevigné (2001), Cook and Ellis (2001), Roweis (2001), and Wang and Brown (1999), and references therein.

4   TWO–CHANNEL SOURCE SEPARATION

We consider the case where mixture signals composed of point source signals and additive background noise are recorded at different microphone locations In most practical situations the recorded microphone signals however contain a significant amount of reverberation. This phenomenon can be again modeled as a data generative model and described in the equation below

$$y_l^i = \sum_m a^i(m)x(m-l) + n^i(l)$$
$$(2)$$

where y denotes the observed data, x is the hidden source, a is the mixing filter, and m is the convolution order and depends on the environment acoustics. An important distinction is made between spatially point sources and distributed background noise. Assuming little reverberation, signals originating from point sources can be viewed as identical when recorded at different microphone locations except for an amplitude factor and a delay (Visser *et al.*, 2003). There are many algorithms that attempt to solve this

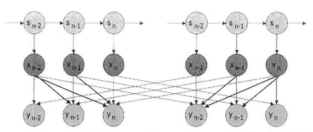

Figure 5.3. A generative model for representing a mixture of audio signals for two sensors. The observations can be modeled in subbands and the source models can be trained on specific audio signals such as the speech signal (Attias, 2003).

multichannel blind deconvolution problem. We outline promising approaches based on viewing this problem in a graphical model.

4.1   Microphone Array Multiple Source Models

In the case of multiple channel observations through an array of micro-phones, the multiple source models can be formulated as follows: $y_l^i = \sum a^i(m)x(m-l) + n^i(l)$, where y^i is the observed signal in channel i, $a^i(m)$ is the mixing filter and n^i is the additive noise signal.

The directed acyclic graph (DAG) model for this mixing problem yields by Attias (2003):

$$p(y|x) = \prod_{ink} N\left(y_{in}[k] | \sum_{jm} H_{ij,m}[k]x_{j,n-m}[k], \lambda_i[k]\right) \tag{3}$$

An EM algorithm estimates the model parameters. As the number of sources increases the E step is computationally intractable and Attias (2003) proposes to use a variational approximation to obtain the posterior distribu-tion. The benefit of solving the multichannel representation is that it not only provides with separated signals but also with the mixing filters, which provide information about the source locations with respect to each other. This addi-tional information is useful in tracking a specific audio signal.

4.2   Separation of Real World Recordings

The separation of real world recordings poses difficult problems in many ways. Although the model in equation 2 may be sufficient, it does not take into account non–stationary issues arising from moving sources and environ-mental dynamics. In some cases however, the environmental setting can be controlled and the proposed solutions apply. The example below shows the separation of two voices recorded live in a conference room during a presen-

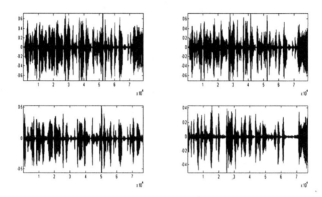

Figure 5.4. At the NSF workshop on speech separation Nov. 2003: Real live recording of two voices (Al Bregman and Te-Won Lee) speaking simultaneously in a conference room environment. The two top plots show the time course of two microphone recordings. Since the microphones were spatially close the plots indicate only minor differences. The 2 bottom plots are the separated voice signals. The signal to noise ratio improvement was about 15dB.

tation. The obtained results are very encouraging and point to the right direction. The audio examples can be found on the author's website.

4.3   Discussion on Multi Channel

There are many algorithms that attempt to solve this multichannel blind deconvolution problem.

Representative work in adaptive signal processing includes Yellin and Weinstein (1996) where higher order statistical information is used to approximate the mutual information among sensory input signals. Extensions of ICA and BSS work to convolutive mixtures include Lambert (1996), Torkkola (1996), Lee *et al.* (1997).

Traditional techniques using microphone arrays include methods for spatial filtering such as beamforming where the time delay between microphones in an array is used to steer a beam towards a sound source and therefore putting a null at the other directions. Beamforming techniques make no assumption on the sound source but assume that the geometry between source and sensors or the sound signal itself is known for the purpose of dereverberating the signal or source localization.

In contrast to beamforming techniques, ICA methods tries to solve the deconvolution and automatic source localization at the same time. The main assumption is statistical independence among sources and it assumes the same number of sources as sensors. Beamforming or blind beamforming (which resembles more the ICA approach) and ICA methods make assumptions in different ways to solve similar problems. Assuming statistical independence among source is a fairly realistic assumption but it comes with several computational constraints such as the required number of sensors to be the same as the number of sources. Furthermore, no sensor noise is usually taken into account.

For practical reasons it is desirable to make stronger assumptions that elevate the sensor number, sensor noise and other constraints. A valid example is the use of a speech model. The characteristics of the speech signal can be included in many ways. In this proposal we plan to elaborate on the source model to include time structure modeled by a hidden Markov model (HMM), the observations in each state are modeled by a mixture of Gaussians in the cepstral domain. This representation is standard for modeling the dynamics of the speech signal. We believe that it will serve as a valid source model. Note that this model is used for the learning of the unmixing filters. The enhanced speech signal is obtained by filtering the observed sensor signals through the unmixing filters.

5   DISCUSSION

Relationship to other methods:

There are several research directions that are related to this research. This work relates to computational approaches for auditory scene analysis (Bregman, 1994, Darvin and Carlyon, 1995, Cooke and Ellis, 2001, deCheveigné, 2001, Wang and Brown, 1999). It also relates to the problem of robustly recognizing words in a realistic noisy environment (Stern *et al.* 1996, Acero, 1990). Computational auditory scene analysis (CASA) techniques focus on techniques for grouping of frequency bands to model the auditory system (Bregman 1994, Darvin and Carlyon 1995, Cooke and Ellis 2001, de Cheveigné, 2001, Wang and Brown 1999). The goal is to model listeners who are adept at extracting sources from mixed sounds although background noise signals can significantly overlap in time and frequency with the target speech signal.

Robust speech recognition in realistic noisy environments can be challenging when the speech signal is mixed with other acoustic sources (Acero, 1990, Stern *et al.*, 1996, Huang *et al.*, 2001). In particular, when two speakers talk simultaneously, most speech recognition systems perform poorly.

6   CONCLUSIONS

We summarized our approaches for separating voices from mixed recordings. In the single channel case, a priori learned basis functions are used to model the temporal structure of the speech signals. A maximum likelihood approach is used to separate a voice from jazz music given only one mixed channel. In case of two microphones, the problem of separating two voices recorded by two microphones has been tackled. The mixing coefficients, time delays and reverberation coefficients are estimated using the maximum likelihood or infomax approach. The two approaches can be combined in a graphical model since both methods can be represented as data generative models where learning involves the representation of signals via the basis functions and inference involves the estimation of sources. The inference part in case of the single channel is nonlinear and linear in the two channel case.

7   ACKNOWLEDGEMENTS

The author would like to acknowledge Hagai Attias, Gil-Jin Jang and Erik Visser for valuable help and fruitful discussions.

References

Acero, A., 1990, *Acoustical and Environmental Robustness for Automatic Speech Recognition*, Ph.D. Thesis, ECE Department, CMU, September 1990.

Attias, H., 1999, Independent factor analysis. *Neural Computation*, **11(4)**:803–852.

Attias, H., 2003, Source separation with a sensor array using graphical models and subband filtering. *Advances in Neural Information Processing Systems 15*, MIT Press, Cambridge

Bell, A.J., and Sejnowski, T.J., 1995, An information–maximization approach to blind separation and blind deconvolution, *Neural Computation* **7**:1129–1159.

Bell, A.J., and Sejnowski, T.J., 1996, Learning the higher–order structure of a natural sound. *Network: Computation in Neural Systems*, **7**:261–266.

Bregman, A.S. 1990, *Auditory scene analysis: The perceptual organization of sound*, MIT Press

Brown, G.J., and Cooke, M., 1994, Computational auditory scene analysis. *Computer speech and language*, **8**:297–336.

de Cheveigné A., 2001, The auditory system as a separation machine, in *Physiological and Psychophysical Bases of Auditory Function*, J. Breebaart, A.J.M. Houtsma, A. Kohlrausch, V.F. Prijs and R. Schoonhoven, eds., Shaker Publishing BV, Maastricht, The Netherlands, pp. 453–460.

Chan, K.-L., Lee, T.-W. and Sejnowski, T.S., 2002, Variational Learning of Clusters of Under-
    complete Nonsymmetric Independent Components, *Journal of Machine Learning
    Research*, **Vol 3**, 99–114.
Comon, P., 1994, Independent component analysis, A new concept? *Signal Processing*,
    36:287–314.
Cooke, M., Ellis, D., 2001, The auditory organization of speech and other sources in listeners
    and computational models, *Speech Communication* **35**, 141–177.
Darwin, C., Carlyon, R., 1995, Auditory grouping, in *The book of perception and Cognition*,
    Vol 6, Hearing, B.C.J. Moore, ed., Academic Press, New York, pp. 387–424.
Ellis, D.P.W., 1994, A computer implementation of psychoacoustic grouping rules. *Proceed-
    ings of the 12th International Conference on Pattern Recognition*, 1994.
Haykin, S., editor, 1994, *Blind Deconvolution*. Prentice Hall, Englewood Cliffs, New Jersey.
Huang, X., Acero, A., and Hon, H.-W., 2001, *Spoken Language Processing*. Prentice Hall,
    Englewood Cliffs,New Jersey.
Hyvärinen, A., Karhunen, J. and Oja, E., 2002, *Independent Component Analysis*, John Wiley
    and Sons.
Jang, G.-J., and Lee, T.-W., 2003, A probabilistic approach to single channel source separation.
    Advances in Neural Information Processing Systems 15, MIT Press, Cambridge.
Jordan, M.I., Ghahramani, Z., and Jaakkola, T.S., 1998, An introduction to variational methods
    for graphical models, in *Learning in Graphical Models*, M. I. Jordan, ed., MIT Press,
    Cambridge, pp. 105–161.
Lambert, R, *Multichannel blind deconvolution: Fir matrix algebra and separation of multipath
    mixtures*. Thesis, University of Southern California, Department of Electrical Engi-
    neering.
Lee, T., Bell, A., and Lambert, R., 1997, Blind separation of convolved and delayed sources, in
    Advances in Neural Information Processing Systems 9. MIT Press, Cambridge.
Lee, T.-W., 1998, *Independent Component Analysis: Theory and Applications*, Kluwer Aca-
    demic Publishers, Boston.
Roweis, S.T., 2001, One microphone source separation. *Advances in Neural Information Pro-
    cessing Systems 13*.
Stern, R.M., Acero, A., Liu, F.-H., and Ohshima, Y., 1996, Signal Processing for Robust
    Speech Recognition, in *Speech Recognition*, C.-H. Lee and F. Soong, eds., Kluwer
    Academic Publishers, Boston, pp. 351–378.
Torkkola, K., 1996, Blind separation of convolved sources based on information maximization.
    IEEE Workshop on Neural Networks for Signal Processing, Kyoto, Japan, **Septem-
    ber 4–6**.
Visser, E., Otsuka, M., Lee, T.-W., 2003, A Spatio–Temporal Speech Enhancement Scheme for
    Robust Speech Recognition in Noisy Environments, **Speech Communications**, **Vol
    41, Issues 2–3**, Pages 393–407.
Wang, D., Brown, G., 1999, Separation of speech from interfering sounds based on oscillatory
    correlation. *IEEE Transactions on Neural Networks* **10** (3), 684–697.
Yellin, D. and Weinstein, E., 1996, Multichannel signal separation: Methods and analysis. *IEEE
    Transactions on Signal Processing*, **44(1)**:106–118.

the beam, the better the ability of the array to select the desired source. The beamwidth and directivity of the delay-and-sum beamformer can be improved by increasing the number of microphones in the array, and by appropriate geometric arrangement of the microphones.

Far more effective than delay-and-sum beamforming is filter-and-sum beamforming. In this method, the signal recorded by each microphone is filtered by an associated filter before the various signals are combined. The spatial characteristics of the beamformer can be controlled by modifying the parameters of the microphone filters.

The design of filter-and-sum beamformers usually involves the estimation of array filter parameters, such that the signal from the desired source is maximally enhanced. Unfortunately, this desired signal cannot be known *a priori*, and the actual design process optimizes alternative criteria that are expected to relate to the enhancement achieved on the desired signal. Sidelobe cancellation techniques design the array filters to attenuate signal energy from directions other than that of the desired source (Griffiths and Jim, 1982). Noise suppression methods design the array to suppress a known or estimated noise (Nordholm *et. al.*, 1999). Least squares methods attempt to maximize the SNR of the array output using estimates of the power spectrum of the desired speech signal (e.g. Aichner *et. al.*, 2003). Thus, effective beamforming requires characterization of either the noise or the desired signal.

Speech recognition systems are repositories of detailed information about the speech signal. They contain statistical characterizations of the spectral measurements for the sounds in a language (usually modelled as hidden Markov models (HMMs)), phonotactic rules for how sounds can follow one another (usually represented as phonetic dictionaries that map words in the language to sequences of phonemes), and statistical or rule–based descriptions of valid word sequences (usually in the form of grammars or N-gram language models). Together these form a complete statistical characterization of every speech signal that represents a valid sentence in the language. Conversely, any valid speech signal can be expected to conform to the statistical characterizations stored in the recognizer.

The beamforming algorithms presented in this chapter are founded on this observation. These algorithms attempt to optimize beamformer parameters such that the signal output by the array maximally conforms to the statistical models stored in an HMM–based speech recognizer (Seltzer, 2003). Specifically, they optimize the filter parameters of a filter-and-sum array to maximize the likelihood attributed to its output by a speech recognizer. Two kinds of beamforming algorithms are presented. The first kind aims to separate out and enhance a speech signal from a mixture of the speech and non-speech signals. Since the interfering signals are not speech and do not con-

Chapter 6

Speech Recognizer Based Maximum Likelihood Beamforming

Bhiksha Raj
Mitsubishi Electric Research Laboratories
bhiksha@merl.com

Michael Seltzer
Microsoft Research
mseltzer@microsoft.com

Manuel Jesus Reyes-Gomez
Columbia University
mjr59@ee.columbia.edu

1   INTRODUCTION

The signal to noise ratio (SNR) of speech signals can be conside
enhanced by recording them through arrays of microphones simultaneo
and combining the recordings properly. The manner in which microp
array recordings must be combined in order to obtain the best results has
the subject of much research over the years.

The simplest array processing method is delay-and-sum beamforr
(Johnson and Dudgeon, 1993). Sounds from any source must travel diffe
distances to the different microphones, the recordings from which are co
quently delayed with respect to each other. In delay-and-sum beamform
the recordings are aligned to cancel out the relative delays of signals from
desired source, and averaged. Interfering noises from sources that are
coincident with the desired source remain misaligned and are attenuatec
the averaging. It can be shown that if the noise signals corrupting the mi
phone channels are uncorrelated to each other and the target speech sig
delay-and-sum beamforming results in a 3 dB increase in the SNR of the
put signal for every doubling of the number of microphones in the array.

The term "beamforming" derives from the fact that such processing ca
shown to selectively pick up signals from a narrow beam of locations aro
the desired source, by attenuating signals from other locations. The narro

form to the models in the speech recognizer, filter parameters can be optimized using the recognizer directly. The second kind of beamforming algorithm attempts to separate signals from multiple speakers who are speaking simultaneously. This is achieved by beamforming separately for each of the speakers: the desired signal for each beamformer is the speech from one of the speakers, while the rest of the speakers are considered as interference. Here an additional complication is introduced by the fact that interfering signals are also speech that may also conform to the models in the recognizer. To account for the multiple conformant signals, the beamforming algorithm utilizes *factorial* hidden Markov models (FHMMs) that are derived by compounding the statistical models stored in the recognizer, to simultaneously model the desired and interfering speech signals. The microphone array filter parameters are estimated such that the likelihood of the output of the array, as measured by the constituent components of the factorial HMM that represent the desired speaker, is maximized.

We note that an HMM–based speech recognition system has two distinct statistical components: acoustic models, that represent statistical constraints on the acoustic manifestation of the speech signal, and a language model that represents linguistic constraints on spoken utterances. For signal enhancement we present two algorithms: one that utilizes only the statistical acoustic constraints and needs deterministic linguistic constraints (Seltzer and Raj, 2001), and a second that utilizes both, statistical acoustic and linguistic constraints (Seltzer *et. al.*, 2002). For speaker separation we present an algorithm utilizes only statistical acoustic constraints and requires deterministic language constraints (Reyes *et. al.*, 2003). The development of speaker separation algorithms that utilize both statistical acoustic and linguistic constraints from the recognizer is left for future work.

2   FILTER-AND-SUM ARRAY PROCESSING

We employ traditional filter-and-sum processing to combine the signals captured by the array. In an optional first step the speech source is localized and the relative delays caused by path length differences from the microphones to the source are resolved, so that the waveforms captured by the individual microphones are aligned with respect to each other. Several algorithms have been proposed in the literature for source localization (e.g. Brandstein and Silverman, 1997), and any of them can be applied here. Source localization and signal alignment is not mandatory, however, the algorithms presented in this chapter have been experimentally verified to work equally well when the signals are not aligned beforehand (Reyes *et. al.*, 2003, Seltzer *et. al.*, 2004).

Once the signals are time aligned, each of the signals is passed through an FIR filter, and the filtered signals are added to obtain the final signal. This procedure can be mathematically represented as follows:

$$y[n] = \sum_{i=1}^{N} h_i[n] \otimes x_i[n - \tau_i] \qquad (6.1)$$

where $x_i[n]$ represents the ith sample of the signal recorded by the ith microphone, τ_i represents the delay introduced into the ith channel to time align it with the other channels, $h_i[k]$ represents the kth coefficient of the FIR filter applied to the signal captured by the ith microphone, $\otimes$ represents the convolution operation, and $y[n]$ represents the ith sample of the final output signal. N is the total number of microphones in the array.

3   BEAMFORMER DESIGN FOR SIGNAL ENHANCEMENT

The beamforming algorithm for signal enhancement attempts to optimize the filter parameters in the filter-and-sum array such that signals from a desired speech source are enhanced in the output of the array. All interfering signals that must be suppressed are assumed to be non-speech signals, or babble-like signals that are not well represented within the speech recognizer.

Figure 6.1 shows the overall design of the filter optimization procedure for signal enhancement. The objective of the algorithm is to choose the filter parameters $h_i[k]$ that maximize the likelihood of $y[n]$, the output of the array, as measured by the recognizer. We distinguish between two versions of the algorithm: a) a calibration algorithm, that utilizes only statistical acoustic

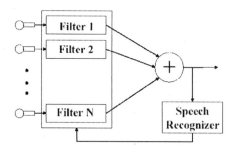

Figure 6.1. Beamformer design for signal enhancement. Filter parameters are set to maximize recognizer likelihood.

constraints from the recognizer and b) an unsupervised algorithm that utilizes both statistical and linguistic constraints.

3.1   Filter Calibration

For the calibration algorithm we assume that the *correct* transcription, *i.e.* the sequence of words in the utterance, is known. Thus the only statistical constraints applied are acoustic. In practice, we utilize only a single *calibration* utterance from the speaker, for which the transcription is known, to optimize the filters. Future utterances by that speaker are processed with the estimated filters. The implicit assumption in this procedure is that speakers do not move too much once their calibration utterance has been recorded. This is not an unrealistic assumption is several situations, such as in automobiles, or users speaking to their desktop computers.

Since the transcription of the calibration utterance is known, an HMM that represents that transcription can now be constructed by concatenating the HMMs for the constituent phonemes that make up the words in the sentence, in appropriate order. We derive the phoneme HMMs from the speech recognizer itself. Filter optimization is then performed using the HMM for the known transcription.

HMM–based speech recognition systems do not operate directly on the speech signal itself. Rather, they operate on a frame–based parameterization of the speech signal. We therefore pose the optimization problem in the context of these frame–based parameterizations. In this chapter we assume that each frame of speech is parameterized as a vector of Mel–frequency cepstral coefficients (MFCC), however, the approach taken is equally applicable to any other type of feature vector. Let $\mathbf{h}$ represent a vector composed of all filter parameters for all microphones. Let $\mathbf{y}_j(\mathbf{h})$ represent the signal $y[n]$ in the jth frame of the calibration utterance, expressed as a function of $\mathbf{h}$. The MFCC vector the jth frame, $\mathbf{z}_j(\mathbf{h})$, is computed as

$$\mathbf{z}_j(\mathbf{h}) = \text{DCT}(\log(\mathbf{M}|\text{DFT}(\mathbf{y}_j(\mathbf{h}))|^2)) \tag{1}$$

where $\mathbf{M}$ represents the matrix of weighting coefficients of the Mel filters. The entire utterance is parameterized into the sequence of vectors $\mathbf{z}_1(\mathbf{h}), \mathbf{z}_2(\mathbf{h}), ..., \mathbf{z}_T(\mathbf{h})$, which we represent as $\mathbf{Z}(\mathbf{h})$.

The likelihood of any utterance must be computed over all possible state sequences through the HMM for the utterance. In order to simplify the computation, we observe that in an HMM–based system, the likelihood of any data sequence is largely represented by the likelihood of the most likely state sequence through the HMMs. The log–likelihood of $\mathbf{Z}(\mathbf{h})$ can therefore be approximated as

$$L(\mathbf{Z}(\mathbf{h})) = \sum_{j=1}^{T} \log(P(\mathbf{z}_j(\mathbf{h})|s_j)) + \log(P(s_1, s_2, ..., s_T)) \qquad (2)$$

where $s_1, s_2, s_3, ..., s_T$ represents the most likely state sequence. $P(\mathbf{z}_j(\mathbf{h})|s_j)$ represents the probability of $\mathbf{z}_j(\mathbf{h})$ computed from the state output distribution of the jth state in this sequence, s_j. $P(s_1, s_2, s_3, ..., s_T)$ is determined by the state transition probabilities of the HMM.

Optimization of $L(\mathbf{Z}(\mathbf{h}))$ requires joint estimation of both $\mathbf{h}$ and the most likely state sequence $s_1, s_2, s_3, ..., s_T$. This can be performed by iteratively estimating the optimal state sequence for a given $\mathbf{h}$ using the Viterbi algorithm, and optimizing $\sum \log(P(\mathbf{z}_j(\mathbf{h})|s_j))$ with respect to $\mathbf{h}$ for that state sequence. $\sum \log(P(\mathbf{z}_j(\mathbf{h})|s_j))$ cannot however be directly optimized and computationally expensive hill–climbing methods must be used to solve for $\mathbf{h}$. To reduce computational effort, we model state output distributions as Gaussians, and assume that to maximize $P(\mathbf{z}_j(\mathbf{h})|s_j)$ it is sufficient to minimize the weighted distance $(\mathbf{z}_j(\mathbf{h}) - \mu_{s_j})^T \mathbf{W}(\mathbf{z}_j(\mathbf{h}) - \mu_{s_j})$ between $\mathbf{z}_j(\mathbf{h})$ and μ_{s_j}, the mean of the output distribution of s_j. Specifically, we assume that the weights matrix $\mathbf{W} = (\text{IDCT})^T (\text{IDCT})$, where IDCT is the inverse discrete cosine transform matrix. This effectively transforms the maximization of $P(\mathbf{z}_j(\mathbf{h})|s_j)$ into the minimization of the Euclidean distance between two log–spectral vectors. Under these assumptions, maximization of $\sum \log(P(\mathbf{z}_j(\mathbf{h})|s_j))$ is equivalent to minimization of the objective function:

$$Q(\mathbf{h}) = \sum_{j=1}^{T} \left\| \text{IDCT}(\mathbf{z}_j(\mathbf{h}) - \mu_{s_j}) \right\|^2 \qquad (6.2)$$

$Q(\mathbf{h})$ can be optimized with respective to $\mathbf{h}$ using hill–climbing methods such as the method of conjugate gradients (Polak, 1971). Details of the gradient derivation for the estimation of $\mathbf{h}$ can be found in Seltzer (2003).

The entire algorithm for optimizing $\mathbf{h}$ from a calibration utterance is thus:

1. Construct an HMM for the transcription of the calibration utterance using HMM components from the speech recognizer.
2. Initialize $\mathbf{h}$ as $h_i[0] = 1/N$, $h_i[k] = 0$, $k \neq 0$.
3. Process the signals using $\mathbf{h}$ to generate an output signal.
4. Determine the optimal state sequence through the HMM for the calibration utterance using the array output.
5. Use the optimal state sequence and (6.2) to estimate $\mathbf{h}$.
6. If $Q(\mathbf{h})$ has not converged, return to step 3.

Note that time alignment of the signals can be performed before step 3 for better initialization of the algorithm, but it is not essential. If the calibration utterance is recorded simultaneously over a close–talking microphone, features derived from this cleaner signal can be used either to determine the optimal state sequence in step 4, or directly in (6.2) instead of the Gaussian mean vectors. Once the filter parameters **h** are derived from the calibration utterance, they are used on newer utterances by the speaker.

3.2   Unsupervised Filter Estimation

In the unsupervised filter estimation algorithm all constraints are statistical. Both the acoustic model and the language model employed by the recognizer are used to guide the beamforming. Thus, the HMM that is used to measure the likelihood of the output of the array is not merely the HMM for the correct transcription for the recorded utterance, but rather represents the entire expected language. Such an HMM must, of necessity, be very large, and the measurement and maximization of the likelihood of an utterance can be arbitrarily complex. As a result, we resort to an iterative algorithm to perform the optimization.

In each iteration, we process the signal using the current estimate of the filter parameters and perform speech recognition on the output of the array. The recognizer's output is a string of words, that is then assumed to be the true transcription for the utterance. An HMM is then constructed for this transcription, and filter parameters are optimized in the manner described in Section 3.1. The entire algorithm for estimating the optimal filters for an utterance can be stated as follows:

1. Initialize **h** as $h_i[0] = 1/N$, $h_i[k] = 0$, $k \neq 0$.
2. Process the signals using **h** to generate an output signal.
3. Perform speech recognition on the array output to obtain a word sequence.
4. Construct an HMM for the recognized word sequence using HMM components from the speech recognizer.
5. Determine the optimal state sequence through the HMM using the current array output.
6. Use the optimal state sequence and (6.2) to estimate **h**.
7. If $Q(\mathbf{h})$ has not converged, return to step 3.

Once again, time alignment of the signals can be performed before step 2 for better initialization of the algorithm, but it is not essential. It is important to note that unlike the calibration algorithm, the unsupervised filter estimation algorithm is applied to every recorded utterance individually. The estimated filters are hence utterance specific, although experimental evidence suggests that they do indeed generalize to other utterances by the same speaker, from the same location.

4   EXPERIMENTAL EVALUATION OF BEAMFORMING FOR SIGNAL ENHANCEMENT

In this section we describe some experiments evaluating the beamforming algorithms presented in Section 3. The experiments were conducted on two microphone array databases recorded at Carnegie Mellon University. The first, database, which we refer to as the "CMU-8" corpus, was collected in the CMU robust speech recognition laboratory. The data were recorded by a linear microphone array with 8 elements spaced 7 cm apart. The array was placed on a desk and speakers were seated directly in front of it at a distance of 1m. Speakers also wore a close–talking microphone during the recording. Each array recording also has a corresponding clean recording obtained from the close–talking microphone. The laboratory had multiple noise sources, including several computer fans and overhead air blowers. The reverberation time of the room was measured to be 240 ms. The average SNR of the recordings is 6.5dB. The corpus contains 140 recordings comprising 14 utterances each from 10 speakers. Utterances consist of strings of keywords, as well as alphanumeric strings, where the user spelled out answers to various census questions, such as name, address, etc.

The second database, which we refer to as the CMU-PDA corpus was collected on a Compaq IPAQ PDA outfitted with four microphones, using a custom–made frame. The microphones were placed at the corners, forming a rectangle around the PDA that was 5.5 cm across and 14.6 cm from top to bottom. Recordings were made in a room containing several items of furniture, three computers, and a printer. Users held the PDA in whichever hand was most comfortable, and read sentences from the Wall Street Journal, that were displayed on the PDA's screen. A total of 8 speakers recorded approximately 40 utterances each for the corpus. The average SNR of the recorded signals was 13 dB. Simultaneous recordings were also obtained over a close–talking microphone worn by the speakers.

Rather than subjective tests or measurements of SNR, we evaluate array performance by comparing the recognition performance of an automatic speech recognition system on the output of the array, to its recognition performance on unprocessed noisy recordings. Superior processing of the array

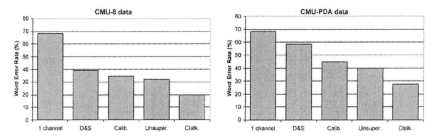

Figure 6.2. Word error rates for the CMU-8 and CMU-PDA microphone array databases. The five bars in each plot show the word error rates obtained from a) signals captured by a single microphone (1channel), b) signals obtained from delay-and-sum array processing (D&S), c) signals obtained from beamformers trained with the calibration algorithm (Calib), d) signals obtained from beamformers trained using the unsupervised algorithm (Unsuper), and e) recordings from a close-talking microphone (clstk).

recordings must result in better recognition performance. In effect, we use the recognizer as a substitute for the human ear that can provide objective measurements[1]. As a comparator, we also report recognition results obtained with simple delay-and-sum beamforming.

The CMU SPHINX-III speech recognition system with context–dependent continuous density HMMs with 8 Gaussian/state, trained using 7000 utterances from the WSJ0 training set was used in all experiments. In all experiments, 20-tap FIR filters were estimated for all microphones. For the calibration experiments one utterance from each speaker was used to estimate filter parameters, and the rest were processed using the estimated filter parameters. In the unsupervised case, filter parameters were estimated afresh for each utterance.

The recognition results for the two databases are shown in Figure 6.2. We observe that signals processed both by the calibration and unsupervised algorithms result in significant improvements over delay and sum processing. Additionally, we observe that the unsupervised algorithm is often more effective than the calibration algorithm, although the calibration algorithm uses deterministic language constraints whereas the unsupervised algorithm uses only statistical constraints. This is attributable to the fact that the unsupervised beamforming is performed individually on every utterance, and thus computes array parameters that are specific to the spatial and frequency characteristics of that utterance. On the other hand, the calibration algorithm

---

[1.] For both automatic speech recognition systems and human beings, improved intelligibility of the speech signal does not necessarily imply improved SNR, or vice versa. For instance, listening studies have shown that signal enhancement techniques that improve the SNR of noisy speech signals often result in a degradation of the intelligibility of the signal, although they do improve the perceptual quality. This is due to the fact that procedures that attenuate noise often also attenuate spectral components of the speech signal along with the noise.

optimizes filters on a calibration utterance and applies it to future utterances from the speaker, and is thus dependent on the similarity of spatial and frequency characteristics of the calibration and test utterances for effectiveness.

On the whole, the experiments indicate that optimizing the beamformer using the detailed statistical information about speech, that is present in a speech recognizer, can result in highly effective beamforming for signal enhancement.

5   BEAMFORMING FOR SPEAKER SEPARATION

The beamforming algorithm for speaker separation addresses the situation where there are multiple speakers talking simultaneously and the array processing scheme must selectively extract the signal from one of the speakers. In this situation, the interfering signals — signals from speakers other than the one we wish to extract — also match the statistical constraints presented by the recognizer. As a consequence, although one may expect the objective function used for filter estimation, *i.e.* the likelihood of the output of the array as measured by the recognizer, to have multiple local optima, one for each speaker, the iterative algorithms presented in Section 3 are usually unable to arrive at these optima.

It thus becomes necessary to explicitly model the fact that there are multiple speech sources that are simultaneously active. Assuming that the multiple sources are independent of each other, the joint probability for the multiple sources is simply the product of the probability distributions of the individual sources. The probability distribution of any single speech source is modelled by a speech recognizer.

Once again, we note that the speech recognizer in fact represents a combination of two independent sets of statistical constraints: acoustic constraints that are modelled by HMMs, and linguistic constraints that are modelled by a grammar or an N-gram language model. The algorithm presented in this chapter derives only acoustic constraints from the recognizer and assumes that all linguistic constraints are deterministic. *i.e.* that the exact word sequence uttered by each of the speakers is known.

From the known word sequences for each speaker we construct an HMM for that word sequence using components from the recognizer. The constructed HMMs represent the probability distribution for the speakers. The joint distribution for all the speakers can be shown to be a cross product of the HMMs for the individual speakers, *i.e.* a factorial HMM, or FHMM.

In an FHMM each state is a composition of one state from the HMMs for each of the speakers, reflecting the fact that the individual speakers may have

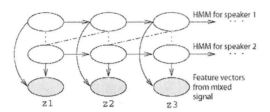

Figure 6.3. The dynamics of a factorial HMM for two speakers. The signal for each speaker follows the dynamics dictated by the HMM for that speaker, independently of the other speaker. The final output, however, is a combination of the outputs of the two HMMs.

been in any of their respective states, and the final output is a combination of the output from these states. Figure 6.3 illustrates the dynamics of an FHMM for two speakers

For simplicity of explanation, we describe the solution to the problem of separating the signals from a speaker when there are only two speakers talking. Extension to the case where there more concurrent speakers is straightforward. Let s_i^1 represent the ith state of the HMM for the first speaker, and s_j^2 represent the jth state of the HMM for the second speaker. Let $s_{i,j}^{1,2}$ represent the factorial state obtained when the HMM for the first speaker is in state i and the HMM for the second speaker is in state j. We assume that the state output densities of all HMM states have parametric forms. Let θ_i^1 and θ_j^2 represent the parameters of the state output densities of s_i^1, and s_j^2 respectively. Let the desired speaker, whose signals we aim to separate, be the kth speaker (where k is either 1 or 2). The output density of $s_{i,j}^{1,2}$ is a function of the parameters of the output densities of its component states:

$$P_k(\mathbf{z}|s_{i,j}^{1,2}) = f_k(\mathbf{z}, \theta_i^1, \theta_j^2) \tag{6.3}$$

where $\mathbf{z}$ represents any feature vector computed from the output of the array, and $f_k(\mathbf{z}, \theta_i^1, \theta_j^2)$ is the actual function of θ_i^1 and θ_j^2 that computes the probability of $\mathbf{z}$ in the factorial state $s_{i,j}^{1,2}$. The subscript k in $P_k(\mathbf{z}|s_{i,j}^{1,2})$ and $f_k(.)$ indicates that these terms are specific to the FHMM that is constructed for the estimation of beamformer parameters for the kth speaker.

The precise nature of the function $f_k(.)$ is unknown. This is because the relative signal levels of the various speakers is unknown, even at the outset. Further, as the algorithm iteratively improves the beamformer for the desired speaker, the levels of the competing speakers in the output of the array is reduced by an unknown degree. At each stage of the algorithm, $f_k(.)$ must reflect the unknown degree of mixing of the various speakers in the current output of the array. As a result of the uncertainty in the initial levels of the speakers or the degree of separation achieved at any iteration, it is difficult, if not impossible, to determine $f_k(.)$ in an unsupervised manner.

We do not attempt to estimate $f_k(.)$. Instead, we begin with the simplifying assumption that the HMMs for the individual speakers have Gaussian state output distributions (in order for this assumption to be valid, the recognizer used must also model HMM states with Gaussians). Additionally, we assume that the state output density for any state $s_{i,j}^{1,2}$ of the FHMM is also a Gaussian whose mean is a linear combination of the means of the state output densities of the component states, s_i^1 and s_j^2.

Let μ_i^1 and μ_j^2 represent the D-dimensional mean vectors of the Gaussian state output density for s_i^1 and s_j^2, respectively, where D is the dimensionality of the feature vectors computed from the output of the array. We define $\mu_{i,j}^{1,2}$, the mean of the Gaussian state output density of $s_{i,j}^{1,2}$, as:

$$\mu_{i,j}^{1,2} = A_k^1 \mu_i^1 + A_k^2 \mu_j^2 \tag{6.4}$$

where A_k^1 and A_k^2 are $D \times D$ weighting matrices. As before, the subscript k indicates that these terms are specific to the FHMM that is constructed for the estimation of beamformer parameters for the kth speaker. We also assume that the Gaussian state output densities of all states of the factorial HMM for the share a common covariance matrix C_k.

The A_k matrices and the covariance matrix C_k are unknown and must be estimated from the current output of the array. The estimation is performed using the expectation maximization (EM) algorithm. In the expectation (E) step of the algorithm, the *a posteriori* probabilities of the various factorial states are found. The factorial HMM has as many states as the product of the number of states in the component HMMs that compose it, and direct computation of the E step is prohibitive. We therefore take a variational approach to the estimation of *a posteriori* probabilities. For further details on the variational estimation of *a posteriori* state probabilities in FHMMs in general, and for the special case of FHMMs for beamforming, we refer the reader to Ghahramani and Jordan (1997) and Reyes *et. al.* (2003).

Once the A_k matrices and the covariance C_k are estimated, all state output densities for the FHMM can be computed. Thus, the FHMM is entirely specified, and filter parameters can be updated. Since the FHMM represents the estimated distribution of a combined signal from multiple speakers, optimizing filter parameters to maximize the likelihood assigned by the FHMM to the output of the array is unlikely to result in a beamformer that selects only the desired speaker. The goal is to optimize filter parameters such that the output of the array most conforms to the HMM for the *desired* speaker, and not to the entire FHMM itself. In order to achieve this, we perform the optimization as follows: we find the most likely state sequence through the FHMM for the current output of the array. Each state in this sequence is a factorial state that represents the compounding of one state each from the HMMs for the individual speakers. The corresponding state sequence for the desired speaker can be obtained by simply identifying the state from the HMM for the desired speaker that contributed to each factorial states in the most likely state sequence. This state sequence represents an estimate for the most likely state sequence through the HMM for the desired speaker. Once this state sequence is obtained, filter parameters are optimized from it using a procedure analogous to that used in Section 3. The overall filter estimation procedure to separate the signals for a desired speaker is as follows:

1. Construct an HMM for the transcription of each speaker using HMM components from the speech recognizer.

2. Construct an FHMM from the HMMs for the individual speakers.

3. Initialize **h** as $h_i[0] = 1/N$, $h_i[k]=0$, $k \neq 0$

4. Process all microphone signals using **h** to generate an output signal.

5. Learn all A_k matrices, and the shared covariance matrix C_k for the FHMM, using the output of the array.

6. Using the learnt A_k and C_k matrices, determine the optimal state sequence through the FHMM using the array output.

7. Extract the state sequence for the desired speaker from the optimal state sequence through the FHMM.

8. Use the optimal state sequence and (6.2) to estimate **h**.

9. If $Q(\mathbf{h})$ has not converged, return to step 3.

The above procedure must be separately performed for each speaker that we wish to separate from the mixed recordings.

6   EXPERIMENTS ON SPEAKER SEPARATION

In this section we present experimental evaluation of the speaker separation algorithm presented in Section 5. Simulated mixed–speaker recordings were generated using utterances from the test set of the Wall Street Journal(WSJ0) corpus. Room simulation impulse response filters were designed for a room of dimensions $4\mathrm{m} \times 5\mathrm{m} \times 3\mathrm{m}$, with a reverberation time of 200msec, using the image method (Allen and Berkley, 1979). The microphone array configuration consisted of 8 microphones placed around an imaginary $0.5\mathrm{m} \times 0.3\mathrm{m}$ flat panel display on one of the walls. Two speakers were situated in different locations in the room and 8-channel recordings were created for the mixtures.

Figure 6.4 shows example waveforms extracted from a mixture of two signals using the proposed algorithm. In this example the signal from the background speaker was scaled to be 20dB below that from the foreground speaker in the mixed signal. The FHMM–based beamforming algorithm is able to extract the signal from the background speaker almost perfectly, except for some scaling and DC shift. The signal for the foreground speaker shows minor residual artifacts from the background speaker in regions where the foreground speaker is silent. However these lie over 23 dB below the overall signal level for the foreground speaker.

Tables 6.1 and 6.2 show a speaker-to-speaker measure for the signals extracted from two different recordings. The reported measure is the ratio of the energy of the signal from the desired speaker to that from the competing speaker in the output of the array, expressed in dB. We refer to this measure as the speaker-to-speaker ratio, or SSR. The higher the value of the SSR, the better the degree of separation of the desired speaker. Table 6.1 shows the results obtained on a recording that contained a foreground speaker and a background speaker who was 20dB below the foreground speaker. Table 6.2 shows the results obtained for a recording where the energy of the signal from both speakers were approximately equal

The tables have three columns. The first column in the tables shows the SSR that can be achieved when the array filters are optimized to minimize the error between the output and the known original unmixed signal for the speaker. This represents an estimate for the best possible separation that can be achieved with a time–invariant beamformer for the signal. The goal of the FHMM–based beamformer is to achieve similar levels of signal separation.

The second column in the tables shows the signal separation achieved using delay-and-sum beamforming. Exact speaker locations and their distances to the microphones are known beforehand in our experiments, and hence the exact relative delays of the signals recorded by the various microphones can be determined. For the delay-and-sum processing reported in the second column of Tables 6.1 and 6.2, signals were aligned perfectly, based on the exact relative delays computed from the known speaker locations. Thus, the results reported in the second column represent the best achievable SSRs with delay-and-sum processing.

The third column in the tables shows the SSRs achieved using the FHMM based beamformer described in Section 5. When the mixed signals have different levels (table 6.1), the FHMM–based beamformer is observed to achieve SSRs comparable with those achieved using perfect knowledge of the unmixed signals. On the background speaker in particular, an impressive SSR

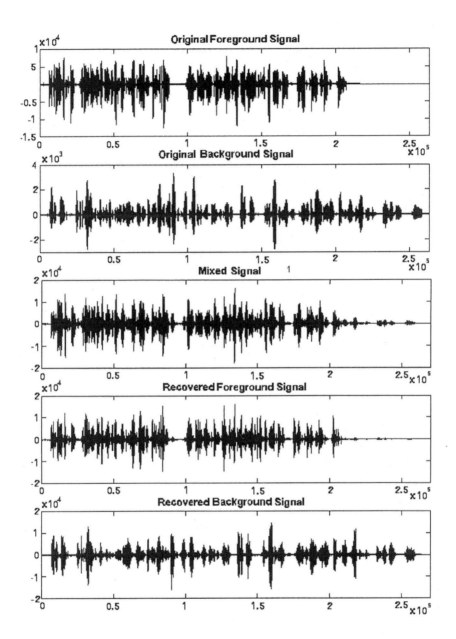

Figure 6.4. Example waveforms for signals extracted from a mixture of two speakers. The top panel shows a mixed signal, and the bottom two panels show the signals extracted for the two speakers. The signal from the background speaker was scaled to be 20dB below the foreground speaker in the mixed signal.

Target Speaker	Clean target	Delay & Sum	FHMM Beamformer
Speaker 1	36dB	-11dB	38dB
Speaker 2	24dB	12dB	23dB

Table 6.1. Speaker-to-speaker ratio (SSR) in dB for a mixture of two speakers where the first speaker is in the background. The table shows the SSR of the signal derived when filters are optimized with full knowledge of the desired clean signal, when simple delay and sum beamforming is used to extract the desired speaker, and the FHMM based beamformer is used to separate signals from individual speakers.

Target Speaker	Clean target	Delay & Sum	FHMM Beamformer
Speaker 1	35dB	1dB	21dB
Speaker 2	19dB	1dB	18dB

Table 6.2. Signal-to-speaker ratio (SSR) of signals derived using three different processing methods on a mixture of two signals of equal energy.

of 38dB is achieved. This is an improvement of nearly 50dB over delay-and-sum processing. When the signal levels of the two mixed signals are comparable (table 6.2), the SSR on the first speakers is somewhat less than that achievable with perfect knowledge of the desired unmixed signal, nevertheless an SSR gain of over 20dB over the original mixed signal is achieved.

7　　CONCLUSIONS AND FUTURE WORK.

In this chapter we have presented microphone array beamforming algorithms that utilize the detailed statistical models in a speech recognizer to selectively enhance a desired speech signal in the presence of speech or non-speech interferences. These algorithms base beamforming on detailed information about the *speech signal*, rather than on any estimate of noise, or the spatial geometry of the recording environment. We show that a conventional HMM–based speech recognition system can provide sufficient statistical constraints on the speech signal in the output of the beamformer to enable us to pick out and enhance a speech signal in a noisy environment.

The beamforming algorithms presented in this chapter have been studied in great detail and have found to be effective on varied data (Reyes *et. al.*, 2003, Seltzer 2003). However, the algorithms remain computationally expensive. Also, the procedures must be performed offline and cannot currently adapt to continuously changing recording environments or speaker location. In addition, the speaker separation algorithm makes the rather serious assumption that word sequences uttered by the speakers are known.

Future work must address the issue of speeding up the computation, and developing online versions that can adapt the beamformer continuously to the incoming signal. The presented speaker separation algorithm only utilizes acoustic constraints from the speech recognizer. Future work must address the development and evaluation of extensions that also incorporate statistical language constraints for speaker separation.

8   ACKNOWLEDGEMENTS

The authors thank Daniel P. W. Ellis and Richard Stern for many useful suggestions. In particular, we thank Dan Ellis for suggesting the SSR measure to evaluate speaker separation performance.

References

Aichner, R. Herbordt, W., Buchner, H., and Kellerman, W., 2003. Least–squared error beamforming using minimum statistics and multichannel frequency domain adaptive filtering. *Proc. Intl. workshop on Acoustic Echo and Noise Control*, Kyoto, Japan, September 2003.

Allen J.B. and Berkley, D.A., 1979. Image method for efficiently simulating small–room acoustics. *Journal of the Acoustic Society of America*, 65:943–950.

Brandstein, M.S. and Silverman, H.F., 1997. A practical methodology for speech source localization with microphone arrays. *Computer Speech and Language*, 11:91–126.

Ghahramani, Z. and Jordan, M.I., 1997. Factorial hidden Markov models. *Machine Learning*, 29:245–275.

Griffiths, L.J. and Jim, C.W., 1982. An Alternative Approach to Linearly Constrained Adaptive Beamforming. *IEEE Trans. on Antennas and Propagation*, 30:27–34.

Johnson, D.H. and Dudgeon, D.E., 1993. *Array Signal Processing: Concepts and Techniques.* Prentice Hall, New Jersey.

Nordholm, S., Clasesson, I., and Dahl, M., 1999. Adaptive microphone array employing calibration signals: an analytical evaluation. *IEEE Trans. on Speech and Audio Processing*, 7:241–252.

Polak, E., 1971. *Computational Methods in Optimization*. Academic Press, New York.

Reyes, M.J., Raj, B., and Ellis, D.P.W., 2003. Multi–channel source separation by factorial HMMs. *Proc. IEEE conf. on Acoutics Speech and Signal Processing* (ICASSP2003), Hong Kong, May 2003.

Seltzer, M.L. and Raj, B., 2001. Calibration of microphone arrays for improved speech recognition. *Proc. European Conference on Speech Communication and Technology (Eurospeech 2001)*, Aalborg, Denmark, September 2001.

Seltzer, M.L., Raj, B. and Stern, R.M., 2002. Speech recognizer based microphone array processing for robust hands–free speech recognition. *Proc. IEEE Intl. Conference on Acoustics, Speech and Signal Processing (ICASSP 2002)*, Orlando Florida, May 2002.

Seltzer, M.L., 2003. *Microphone array processing for robust speech recognition*. Ph.D dissertation, Department of Electrical and Computer Engineering, Carnegie Mellon University, July 2003.

Seltzer, M.L., Raj, B. and Stern, R.M., 2004. Likelihood maximizing beamforming for robust hands–free speech recognition. *IEEE trans. on Speech and Audio Processing*. To appear in 2004.

Chapter 7

Exploiting Redundancy to Construct Listening Systems

Paris Smaragdis
Mitsubishi Electric Research Laboratories
paris@merl.com

1   INTRODUCTION

It is no revelation that the world we live in is dominated by statistical regularities. Most cases of informative sensory input exhibit a strong sense of structure. This fact has been used by many researchers in the field perception to formulate theories on the importance of this structure. In this chapter we will be highlighting the importance of using auditory structure to construct artificial listening systems, especially when we are interested in source separation.

Starting with Helmholtz at the beginning of the 20th century, researchers and thinkers came to recognize the effect of the environment on the development of our perception. A major step was taken by Barlow (Barlow 1961, Barlow 1989), who was the first to clearly state that the statistical nature of sensory messages is important to all levels of perception ranging from the sensory receptors to higher cognitive levels. He was particularly interested in redundancy and how it can be taken advantage of. Later reformulations of his work suggested that factorial coding, in which sensory input is broken up in statistically independent elements, was a function of the perceptual system. In the last 15 years this field has gathered momentum and we've seen many important publications relating sensory statistics to perception. The most recent work in this field comes from the Independent Component Analysis (ICA) front, in which factorial coding optimization was used to self-organize perceptual mechanisms from the statistics of natural sensory signals (Olshausen and Field 1996). In this report we will show some of these results in the auditory setting.

The basic idea behind all that work is simple. Raw sensory input requires very large bandwidth and includes a lot of redundant information. This is because all natural sensory input is full of repetitions of the same elements (edges and lines for images, harmonic ratios for natural sounds, etc.). A lot of

this redundant information has been 'understood' by our perceptual development and has influenced the evolution our sensory systems. As a result,
throughout our sensory systems we find detectors tuned to these features. In
this remaining sections we will show how using this notion of redundancy
reduction can computationally evolve a lot of the auditory perception
elements.

2   TOOLS FOR REDUNDANCY REDUCTION

Although there are numerous ways to achieve redundancy reduction, there
are two algorithms that have proved to be most useful in the context of audio
analysis, Principal Component Analysis (PCA) and Independent Component
Analysis (ICA). Although the results in the subsequent sections can be repli-
cated using a variety of other tools, due to their relative simplicity for the
remainder of this chapter we will be primarily using these two methods. A
short introduction of these follows.

2.1   PRINCIPAL COMPONENTS ANALYSIS

Principal Component Analysis (PCA) is one of the most well known methods for data analysis. It has been around for many years and has come to be
known by many different names in various disciplines (Karhunen-Loève or
Hotelling transform). The objective of this method is dual. Primarily it is to
massage input data in a form where the multiple dimensions of the input data
are mutually decorrelated. As a side effect the significance of each decorrelated dimension, in terms of energy contribution in the original input, is also
estimated. Upon concluding the computations the user is presented with a
ranking on which dimensions are most important and based on this ranking
we can throw away non-important dimensions and reduce the dimensionality
of the data with minimal side effects.

Computation is fairly straightforward and is based on second–order statistics of the data. Having a set of N-dimensional data in a random vector $\mathbf{x}$ we
obtain their covariance matrix $\mathbf{C_x} = \mathrm{E}\{\mathbf{x} \cdot \mathbf{x}^T\}$, where $\mathrm{E}\{\cdot\}$ denotes expectation.
We extract the eigenvectors of $\mathbf{C_x}$ as columns in a matrix $\mathbf{V}$ and its eigenval-
ues as diagonal entries in a matrix Λ. The matrix $\mathbf{V}$ is a linear transform that
can decorrelate $\mathbf{x}$, that is if we set $\mathbf{y} = \mathbf{V}^T \cdot \mathbf{x}$ then its covariance $\mathbf{C_y} = \mathrm{E}\{\mathbf{y} \cdot \mathbf{y}^T\}$ is
diagonal. This means that there are zero correlations between different dimen-
sions of $\mathbf{y}$, therefore no redundancy in the signal as far as correlation goes. If
dimensionality reduction is needed then by noting the smallest valued eigen-
values in Λ we can omit the corresponding eigenvector columns in $\mathbf{V}$ and use

the remaining matrix for the transformation. This results in a **y** which is still decorrelated, but with lesser dimensions than the starting input.

More information about PCA and details on the computations as well as alternative algorithms can be found in Jolliffe (1986), Roweis (1998) and Haykin (1994).

2.2 INDEPENDENT COMPONENT ANALYSIS

Independent Component Analysis (ICA) can be seen as a generalization of PCA. PCA only achieves redundancy reduction by eliminating correlations between different dimensions. Correlations are 2nd order statistics, and in general audio signals exhibit higher–order statistics as well. In principle, ICA goes the extra step and performs the same process as PCA for all orders of statistics. Although it might not sound like it, this makes an enormous difference in results. When it comes to audio processing decorrelation means nothing to our ears, but independence (the more stringent condition that ICA imposes), has a lot of semantic meanings, some of which we'll see later in this chapter.

Unfortunately general and complete solutions to ICA are a computationally impossible task since one could be facing infinite orders so various shortcuts to getting a viable solution have been proposed. A couple of interpretations of independence which are perhaps more relevant to PCA are the following. Independence can be loosely defined as decorrelation between all possible transformations of our data. Given out data **x** we know that its dimensions are mutually independent if $E\{g(\mathbf{x}) \cdot f(\mathbf{x}^T)\}$ for all measurable $g(\cdot)$ and $f(\cdot)$ (Hyvärinen 1999). This is again a problem with infinite constraints, however by carefully selecting the two functions $g(\cdot)$ and $f(\cdot)$ one can perform ICA using only a couple of their instances. A more algebraic approach is by performing the equivalent of PCA for 4th order statistics which in conjunction with PCA is often enough to do the job. In the 4th order we are now facing the equivalent of the covariance matrix, which is the quadricovariance tensor. Diagonalizing this entity is much more complicated than performing eigenanalysis, it is however possible (Cardoso 1990) and often a good approximation for ICA. More details about ICA and available algorithms are available in Hyvärinen (1999).

3 AUDITORY PREPROCESSING

The use of harmonic decompositions for auditory processing has had a long history and many successful applications. However the reason for using such transforms has never been clearly justified, neither in the computational nor on the perceptual domain. In this section we will present some evidence as to why localized sinusoids are good bases for auditory analysis and we'll relate that to the development of the cochlea and the evolution of computational auditory decomposition algorithms ranging from the DCT to the Gabor wavelets.

The use of harmonic basis functions came to light with the introduction of the Fourier transform by Jean-Baptiste Joseph Fourier, early in the 19th century. After the introduction of the FFT algorithm in the 1950's and further work on the statistical significance on the DCT, harmonic analysis decompositions became very popular and were shortly expanded so as to address many computational issues. By now we have seen the field of time/frequency analysis burgeon providing a very rich selection of choices satisfying various conditions that often arise in auditory modeling. In our work we set forth to address the question of why these harmonic transforms are so useful. We will not do this using mathematical analysis (as has been done by many in the past), but rather from an evolutionary audition perspective. We will base our work on a redundancy reduction principle that has been the staple of many perceptual evolution theories, and we will attempt to "evolve a cochlea" based on exposure to natural sounds.

The framework of our work is based on a simple rule, we seek to find a linear transformation (such as PCA and ICA) that will attempt to produce a non-redundant projection of incoming auditory data. To define the process more concretely, we will start with a 'training' sound $s(t)$. In order to be able to impose a linear transform on it we will segment it as a set of time windows using the vector formulation $\mathbf{s}(t) = [s(t+1) \ldots s(t+N)]^T$. In which case N is an integer representing the length of our windows. In this form we can represent the linear transform we are about to perform as a matrix multiplication, $\mathbf{x}(t) = \mathbf{W} \cdot \mathbf{s}(t)$. Where $\mathbf{W}$ is a $N{\times}N$ matrix. Had this matrix been a DFT or DCT matrix this would have been the respective transform for each window. For our interests we want to find a custom transform $\mathbf{W}$ to remove redundancy from the data $\mathbf{x}(t)$. This can be achieved in various ways, we choose to employ an Independent Component Analysis (ICA) algorithm to do this.

For our experiments we used fifty single sentence samples from the TIMIT speech database as the sound S, and employed the Amari algorithm (Amari *et al* 1996), to extract to perform ICA. Numerical simulations provided us with the desired matrix/transformation $\mathbf{W}$. The columns of this matrix $\mathbf{W}$, will contain the auditory components (or basis functions) that decompose the incoming sound data in this sparse way. In Figure 7.1 we provide time plots of some basis functions, and the energy of all the bases on the time/frequency plane. It is clear to see that the components are localized sinusoids (localized in both time and frequency), which also exhibit a time frequency trade-off akin to a Gabor transform. The important point to make here is that without any external hints we have deduced that in statistical terms time–localized sinusoids with time/frequency trade-off behavior are optimal in terms of removing redundancy across audio data. This fact makes them an optimal feature decomposition for extracting any information from an input sound.

Similar work has been done in deriving the DFT and DCT family transforms from PCA, however no time localization or time/frequency trade-off were observed in these cases which provided long bases covering the entire input window (Ahmed 1974, Rao and Yip 1990).

The importance of these results is twofold, first of all we are provided with experimental proof as to why time localized harmonic representations with the appropriate time/frequency trade-off are useful for sound analysis. On the other hand we present a procedure that adheres to modern theories on computational and neurobiological perception, where the goal of the perceptual system is assumed to be that of a redundancy–reducing machine. Based on these theories we can make an argument for the evolution of the cochlea as a frequency transform, being guided by such a principle. Similar work on the visual domain has yielded basis functions similar to the ones measured on the retina, and the V1 and V2 brain regions (Deco and Obradovic 1996, Olshausen and Field 1996). What is interesting here is that by using other types of stimuli we can obtain bases that are suited for these signals. For example using music an the input we obtain longer in time bases, which mirror the sustaining note aspects of musical notes, using natural sounds, which have a shorter time scale than speech or music, results in more narrow in time sinusoids. Likewise the frequency content of the input signal is also somewhat mirrored by the density of the recovered bases in frequency.

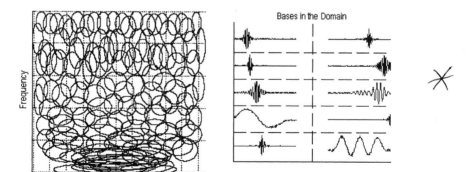

Figure 7.1. Plots of ICA bases. In the left figure we see the time plots of randomly selected bases. Although approximate, the bases clearly are based on localized sinusoids of varying frequencies and time positions. In the right figure we see the energy of the bases in the time/frequency plane. Note how the bases of the lower frequencies have a different time/frequency spread ratio from the higher frequency bases.

4   GROUPING

Grouping of basic elements, such as the ones we derived above, is often used as a natural next step to performing source separation or scene analysis. The input sound is broken down by a basis decomposition, and the resulting elements can be separated and picked according to preset rules so as to reconstruct individual parts of the scene. Unfortunately for the engineering–minded, the rules that we often find to dictate proper grouping are heuristic in nature and hard to implement or to describe mathematically. In this section we will describe a more mathematically precise way to deal with grouping.

As we did before we will once more try to exploit the statistical structure of sounds. This time we will explore some of the basic grouping rules and show how they can be unified under one principle. To do so we will use the notion of redundancy once again. As described above, redundancy is a measure that can mirror the amount of structure in data so that more redundancy implies more structure, and vice-versa. Redundancy can be measured in terms of mutual information, a statistical measure that describes how much information is shared between multiple signals. Using mutual information as an indication of structure and redundancy we can set up a set of experiments on auditory grouping.

We set up an experiment as follows, we construct a simple scene describing an auditory grouping criterion. For example consider the scene: $s(n,t) =$ $[\cos(t) \cos(n{\cdot}t)]$, it contains two sinusoids, with a frequency ratio of n. We change the parameter n and for each value we measure the mutual information between the two resulting sinusoids. From our knowledge of auditory grouping we know that if two sinusoids are harmonically related they tend to fuse as one sound. So if mutual information was a good way to judge grouping, we should see some unique behavior for integer values of n. And so is the case by observing Figure 7.2. Whenever mutual information peaks we have a case where grouping is more likely to occur. We also demonstrate selected examples of other forms of grouping in Figures 7.3, 7.4 and 7.5. In all cases we see a strong correlation between mutual information peaking and psychoacoustical expectation for fusion, which leads us to believe that mutual information is a good cost function by which to evaluate grouping.

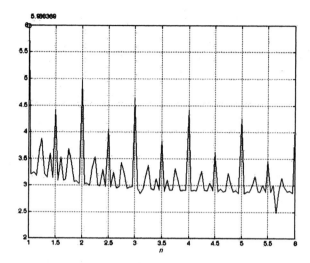

Figure 7.2. Mutual information as a function of harmonicity between two sinusoids. Note that it peaks for integer values of frequency ratios, which are the points where auditory grouping is most likely to occur. Intermediate peaks correspond to the case where the two sinusoids assume the ratios between two higher harmonics of an assumed sound. Also note how mutual information drops as the tones are still harmonic but by a large ratio. Inserting missing harmonics boosts mutual information in that case.

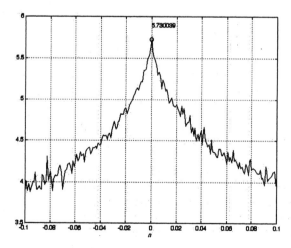

Figure 7.3. Mutual information as a function of common amplitude modulation. Two amplitude modulated tones were used, the parameter n is a distance measuring their modulation dissimilarity. For n = 0 the modulations were identical, and became more different with increasing magnitudes of n. Note how the mutual information peaks at the common modulation.

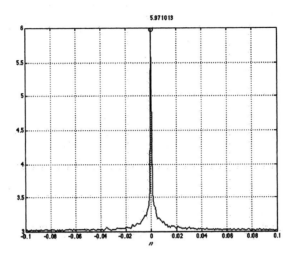

Figure 7.4. Just as the example in Figure 7.3 this time we measure frequency modulation. The parameter n is again a distance between the two modulating functions. At the point where the modulations become identical and the tones tend to fuse we have a mutual information peak.

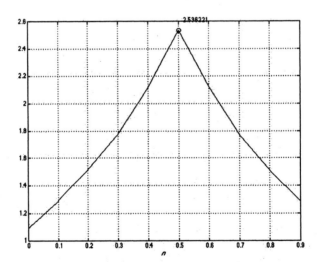

Figure 7.5. Mutual information measurements as a function of common onset. Two tones were used, one starting at time 0.5. The parameter n is the starting time of the second tone. The mutual information peak is at the point where the two tones start at the same time.

This observed relation between signal redundancy and grouping rules can be used to our advantage by constructing algorithms that can perform grouping by minimizing mutual information. One such example is the family of ICA algorithms, which can be easily adapted to perform grouping (Smaragdis 2001). Although this presents a more elegant methodology for grouping, the really interesting fact is that this is the same process that led us to sinusoidal bases in the previous section. By observation of the statistics of sound and by application of the same algorithms we have automatically deduced known information about two major functions about our auditory system.

5   SCENE ANALYSIS

Preprocessing and grouping are often used as first stages for scene analysis. So far we demonstrated how we can take advantage of the regularity of sound to perform these first stages, now we will demonstrate how we can do the same thing again for scene analysis.

Scene analysis is a very hard problem, and the reason it is so is mostly because it is usually ill-defined. In this work we will define scene analysis as the problem of identifying sources in a scene and being able to pinpoint them in time and by their spectral shape. As a side–effect we will have the ability to reconstruct individual sources, this however is not the primary goal. This is a simplified model which does not take into account a lot of the temporal structure, but as we'll demonstrate that it is capable of performing very well on real–world scenes.

Once again we will use the notion of statistical regularity to extract the information we want. We start from a magnitude spectrogram $\mathbf{F}$ of a scene $s(t)$. We have $\mathbf{f}(t) = \| \mathrm{DFT}([s(t + 1) \ldots s(t + M)]) \|$, and we can 'collect' all $\mathbf{f}(t)$, in an $M{\times}N$ matrix $\mathbf{F}$ so that $\mathbf{F} = [\mathbf{f}(1) \ldots \mathbf{f}(N)]$. This time we will apply PCA on $\mathbf{F}$, to reduce the number of dimensions to K, and then we will apply ICA to force these K dimensions to be independent. The entire process can be summarized by one linear transformation $\mathbf{A}$, such that the independent data $\mathbf{H}$ will result from $\mathbf{H} = \mathbf{A} \cdot \mathbf{F}$. With some manipulation we rewrite this as $\mathbf{F} = \mathbf{A}^{-1} \cdot \mathbf{H} = \mathbf{W} \cdot \mathbf{H}$, where $\mathbf{W} = \mathbf{A}^{-1}$. Given that the input data is non-negative (being magnitude spectra), we could also employ non-negative ICA (Plumbley 2002), although it is not strictly necessary (Cichocki 2003).

Examining the product $\mathbf{W} \cdot \mathbf{H}$ we can interpret it as a components–based synthesis. Each column of $\mathbf{W}$ is multiplied with each corresponding row of $\mathbf{H}$ to produce a section of the spectrogram $\mathbf{F}$. The sum of these sections will define the overall approximation. If $K = M$, then the approximation will be perfect, if $K \ll M$, then we see interesting behavior where the components in

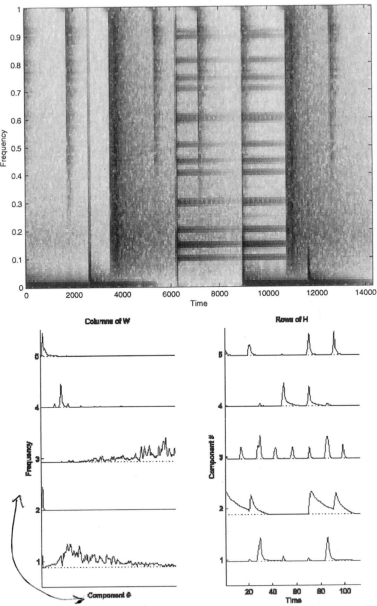

Figure 7.6. In the top figure we can see a spectrogram of a drums bar. We can visually see four instances of a bass drum at the low frequencies, two instances of a snare drum, two instances of a harmonic-like instrument (a cowbell), and some high–frequency instances of the hi-hat. The bottom figure displays a component decomposition of the spectrogram. Note how the elements that composed the original scene have been approximated spectrally by the columns of **W** and temporally by the rows of **H** (also note that the left plot is in log scale for better visualization, but the extracted component spectra on the right are plotted in linear scale).

the columns of **W** and the rows of **H** describe individual sources present in the spectrogram. Consider the spectrogram scene in Figure 7.6, being the analysis of a bar of drumming. Upon visual examination we can see that it is composed of a variety of sources, each of which is a different drum. There is overlap between the sources in the frequency and the time axes, so clean separation is impossible. If we assume however that the resulting spectrogram is an addition of a set of spectrograms, each describing an individual source we can model it using the aforementioned method. The results are shown in Figure 7.6.

Given that this type of decomposition can give us a spectral and temporal description of the components that make up the input scene, we can attempt to perform reconstruction by approximating the input spectrogram using only one component at a time. This can easily be done by the multiplication of the ith column of **W** with the ith row of **H** to extract the magnitude spectrogram of the ith component. Figure 7.7 demonstrates the individual components reconstruction for the scene in Figure 7.6. To make an audible reconstruction we can use the phase of the original spectrogram and mask it by the amplitude masks that are displayed in Figure 7.7.

Similarly we can analyze more complex scenes and derive multiple components that construct the original input in terms of sources. The more the number of components we request (in the form of the parameter K) the more detailed the definition of the source becomes. If for example we had asked for $K=100$ components in the above example we would be presented with elements that make up what we perceive to be sources in the scene. Since defining a source exactly is the product of subjective judging it is hard to say what an optimal setting for K would be.

In the case of speech signals this approach would extract features that correspond to individual phonemes rather than the entire spoken content (this is a side effect of the fact that we do not account for temporal continuity, although it can be statistically argued that the sources are indeed the phonemes, and not the entire spoken content). Piecing these components together it is possible to reconstruct extracted speech.

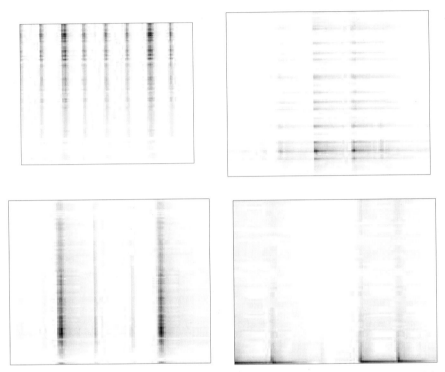

Figure 7.7. Spectrograms of the extracted components. Clockwise from top left, the high-hat, the cowbell, the snare drum and the bass drum.

6   CONCLUSIONS

This report highlighted some of the intricate relationships between the statistics of natural sounds and auditory processing. We have argued that many of the common steps that we often take to perform computational audition can be seen as processes driven by the nature of sound, and not so as steps inspired by human physiology or engineering. We have shown how different aspects of hearing can be explained using a simple and common rule exploiting the statistical structure of sound. Although the methods we employed are very simple, the results are just as promising as using any other more complex approach. We hope that the simplicity and the elegance of this approach will inspire further work along these lines, and give rise to more investigations in the field of computationally evolving audition.

References

Amari, S-I., Cichocki, A., and Yang, H.H., 1996, A new learning algorithm for blind signal separation, *Neural Information Processing Systems*, 1996.

Ahmed, N., Natarajan, T., and Rao, K.R., 1974, Discrete Cosine Transform, *IEEE Transactions on Computers*.

Barlow, H.B., 1961, Possible principles underlying the transformation of sensory messages. In *Sensory Communication,* MIT Press.

Barlow, H.B., 1989, Unsupervised learning, in *Neural Computation* 1, MIT Press, Cambridge MA.

Cardoso, J.-F., 1990, Eigen–structure of the fourth–order cumulant tensor with application to the blind source separation problem. *Proc. ICASSP'90*, pages 2655–2658

Cichocki, A. and Georgiev, P., 2003, Blind source separation algorithms with matrix constraints, *IEICE Trans. Fundamentals*, **Vol. E86–A**

Deco, G. and Obradovic, D., 1996, *An information theoretic approach to neural computing.* Springer-Verlag, New York, New York.

Haykin, S.S., 1994, *Neural Networks: A Comprehensive Foundation*, Macmillan.

Hyvärinen, A., 1999, Survey on Independent Component Analysis, *Neural Computer Surveys*, **Vol. 2**, pp. 94–128.

Jolliffe, I.T., 1986, *Principal Component Analysis.* Springer-Verlag, New York, New York.

Olshausen, B.A., and Field, D.J., 1996, Emergence of simple–cell receptive field properties by learning a sparse code for natural images, *Nature*, **381**.

Plumbley, M.D., 2002, Conditions for non-negative independent component analysis, *IEEE Signal Processing Letters*, **9(6)**, pp177 –180.

Rao, K. and Yip, P., 1990, *Discrete Cosine Transform, Algorithms, Advantages, Applications.* Academic Press.

Roweis, S., 1998, EM Algorithms for PCA and SPCA, *Neural Information Processing Systems* **10**.

Smaragdis, P., 2001, *Redundancy Reduction for Computational Audition: A Unifying Approach.* MIT Doctoral dissertation. Χ

Chapter 8

Automatic Speech Processing by Inference in Generative Models

Sam T. Roweis
Department of Computer Science, University of Toronto
roweis@cs.toronto.edu

1   INTRODUCTION

Say you want to perform some complex speech processing task. How should you develop the algorithm that you eventually use? Traditionally, you combine inspiration, carefully examination of previous work, and arduous trial-and-error to invent a sequence of operations to apply to the input waveform or short–time spectral representation. But there is another approach: dream up a "generative model" – a probabilistic machine which outputs data in the same form as your data – in which the key quantities that you would eventually like to compute appear as hidden (latent) variables. Now perform *inference* in this model, estimating the hidden quantities. In doing so, the rules of probability will derive for you, automatically, a signal processing algorithm. While inference is well known to the speech community as a decoding step for HMMs, exactly the same type of calculation can be performed in many other models not related to recognition.

This chapter explores the use of inference in three separate models, and shows how it can be used to perform surprisingly complex speech processing tasks including denoising, source separation, pitch tracking, timescale modification and estimation of articulatory movements from audio.

2   A FACTORIAL–MAX MODEL OF LOG SPECTROGRAMS

In this section, we review the astonishing max approximation to log spectrograms of mixtures, show why this motivates a "reflltering" approach to separation and denoising, and then describe how the process of inference in factorial probabilistic models performs a computation useful for deriving the

masking signals needed in refiltering. A particularly simple model, Factorial–
Max Vector Quantization (MAXVQ), is introduced along with a branch-and-
bound technique for efficient exact inference and applied to both additive
denoising and monaural separation. Our approach represents a return to the
ideas of Ephraim, Varga and Moore (Varga and Moore, 1990) but applied to
auditory scene analysis rather than to speech recognition.

2.1   Sparsity and Redundancy in Spectrograms

The *sparse* nature of the speech code across time and frequency is one of
the key features exploited by many successful speech processing algorithms.
Roughly speaking, most low noise narrow frequency bands carry substantial
energy only a small fraction of the time and thus it is rare that two independent sources inject large amounts of energy into the same subband at the same
time. (Figure 8.1b shows a plot of the relative energy of two simultaneous
talkers in a narrow subband, most of the time at least one of the two sources
shows negligible power.)

The speech code is also *redundant* across time–frequency. Different frequency bands carry, to a certain extent, independent information and so if
information in some bands is suppressed or masked, even for significant dura-
tions, other bands can fill in. (A similar effect occurs over time: if brief
sections of the signal are obscured, even across all bands, the speech is still
intelligible, while also useful, we do not exploit this here.) This is partly why
humans perform so well on many monaural speech separation and denoising
tasks. When we solve the cocktail party problem or recognize degraded
speech, we are doing structural analysis, or a kind of "perceptual grouping"

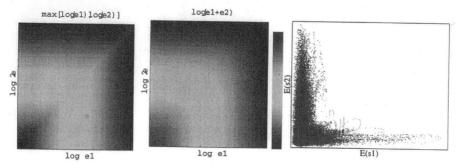

Figure 8.1. (left) Relationship between log of sum and max of logs, each function's value is
shown using the color scale indicated in the middle. Significant differences occur only when
$e_1 \approx e_2$ and both are large. (right) Relative energy of two sources in a single subband, few
points appear on the diagonal.

on the incoming sound. There is substantial evidence that the appropriate sub-
parts of an audio signal for use in grouping may be narrow frequency bands
over short times.

2.1.1   The Log–Max Approximation.

When two clean speech signals are mixed additively in the time domain,
what is the relationship between the individual log spectrograms of the
sources and the log spectrogram of the mixture? Unless the sources are highly
dependent (synchronized), the log spectrogram of the mixture is almost
exactly the *maximum* of the individual log spectrograms, with the maximum
operating over small time–frequency regions (fig. 8.2). This amazing fact,
first noted by Roger Moore in 1983, comes from the fact that unless e_1 and e_2
are both large and almost equal, $\log(e_1 + e_2) \approx \max(\log e_1, \log e_2)$ (fig. 8.1a).

The sparsity of the speech code is what makes this approximation useful in
practice, since the approximation only breaks down when two sources put a
large amount of energy into the same narrow frequency band at the same time,
which rarely occurs.

2.1.2   Masking and Refiltering.

To exploit the redundancy of the speech code, we will focus on narrow
frequency bands over short times and try to group these \subparts" of the signal together, based on whether they belong to the same source or not. If we
can collect enough parts that we are confident belong together, we can discard
the rest of the signal and recover the original source based only on the
grouped parts.

To generate these parts computationally, we can perform multiband analysis – break the original speech signal $y(t)$ into many subband signals $b_i(t)$
each filtered to contain only energy from a small portion of the spectrum. The
results of such an analysis are often displayed as a spectrogram which shows
log energy (using color or grayscale) as a function of time and frequency.
(Think of a spectrogram like a musical score in which the color or grey level
of the each note tells you how hard to hit the piano key.)

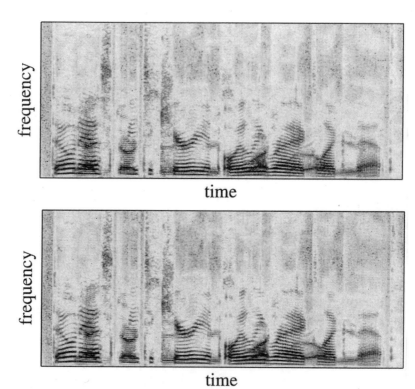

Figure 8.2. (top) Log spectrogram of amixture of two sources. (bottom) Elementwise maximum (within each time–frequency b in) of log spectrograms of original sources. The two spectrograms are almost identical, although the bottom one is an approximation of what the top one ought to looklike based on a very simple combination model.

The basic idea of *refiltering* (Roweis, 2001, Green *et al.*, 2001) is to separate or denoise sources by selectively reweighting the $b_i(t)$ obtained from multiband analysis of the original mixed or corrupted recording. Crucially, unlike in unmixing algorithms, the reweighting is not constant over time, it is controlled by a set of masking signals. Given a set of masking signals, denoted $\alpha_i(t)$, a clean source $s(t)$ can be recovered by modulating the corresponding subband signals from the original input and summing:

$$\underbrace{s(t)}_{\text{estimated source}} = \overbrace{\alpha_1(t)}^{\text{mask 1}} \underbrace{b_1(t)}_{\text{sub-band 1}} + \ldots + \overbrace{\alpha_K(t)}^{\text{mask K}} \underbrace{b_K(t)}_{\text{sub-band K}} \quad (1.1)$$

The $\alpha_i(t)$ are gain knobs on each subband that we can twist over time to bring bands in and out of the source as needed. This performs masking on the original spectrogram.[1] This approach, illustrated in figure 8.3, forms the basis of many CASA systems (Green *et al.*, 2001, Brown and Cooke, 1994). The basic intuition is to "gate in" subbands deemed to have high signal to noise and to be part of the source we are trying to separate and "gate out" subbands when they are deemed to be noisy or part of another source.

For any specific choice of masking signals $\alpha_i(t)$, refiltering attempts to isolate a *single clean source* from the input signal and suppress all other sources and background noises. Different sources can be isolated by choosing a different set of masking signals. Although, in general, masking signals are real–valued, positive quantities that may take on values greater than unity, in practice the (strong) simplifying assumption that $\alpha_i(t)$ are binary and constant over a timescale τ of roughly 30ms can be made. This assumption is physically unrealistic, because the energy in each small region of time–frequency never comes entirely from a single source. However, for small numbers of sources, this approximation works quite well (Roweis, 2001), in part because of the effect illustrated in figure 8.1b. (Think of ignoring collisions by assuming separate piano players do not often hit the same note at the same time.) (Refiltering can also be thought of as a highly nonstationary Wiener filter in which both the signal and noise spectra are re-estimated at a rate $1/\tau$, the binary assumption is equivalent to assuming that over a timescale τ the signal and noise spectra are nonoverlapping.) It is a fortunate empirical fact that refiltering, even with piecewise constant binary masking signals, *can* cleanly separate sources from a single mixed recording.[2]

2.2 Multiband grouping as a statistical pattern recognition problem

Since refiltering for separation and denoising is indeed possible if the masking signals are well chosen, the essential statistical problem is: how can the $\alpha_i(t)$ be computed automatically from a single input recording? The goal is to group together regions of the spectrogram that belong to the same auditory object (and have high signal-to-noise). Fortunately, natural auditory signals – especially speech – exhibit a lot of regularity in the way energy is distributed across the time–frequency plane. Grouping cues based on these regularities have been studied by psychophysicists and are hand built into many CASA systems. Cues are based on the idea of suspicious coincidences: roughly, "things that move together likely belong together". Thus, frequencies which exhibit common onsets, offsets, or upward/downward sweeps are more likely to be grouped into the same stream. Also, many real world sounds have harmonic spectra, so frequencies which lie exactly on a harmonic "stack" are often perceptually grouped together. (Musically, piano players do not hit keys randomly, but instead use chords and repeated melodies.) The approach advocated here to use statistical learning methods to discover these regularities from a large amount of speech data and then to use the learned models to compute the masking signals for new signals in order to perform refiltering.

2.2.3   MAXVQ: Factorial–Max Vector Quantization.

It is often advantageous to model complicated sensory observations using a number of separate but interacting causes. One general way to pursue this modeling idea is to have a fixed number M of vector quantizers (or mixture models), each of which proposes an output, and then have some way of combining the output proposals into a final observation. We can think of this as a bank of quantizers, which feed their chosen prototypes into a nonlinear combination box that computes the final output.[3]

Motivated by the observation above regarding the max approximation to log spectrograms of mixtures, we propose such a model, called Factorial–Max Vector Quantization (MAXVQ), which uses the Max operation to combine outputs from the various causes. The model has a bank of M independent vector quantizers, each of which stochastically selects a prototype with which to model the observation vector. The final output vector is a noisy composite of the set of proposed prototypes, obtained by taking the *elementwise maximum* of the set and adding nonnegative noise. This generative model is illustrated in figure 8.4.

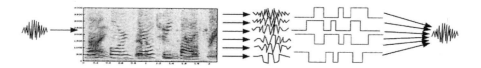

Figure 8.3. Refiltering for separation and denoising. Multiband analysis of the original signal $y(t)$ gives sub-band signals $b_i(t)$ which are modulated by masking signals $\alpha_i(t)$ (binary or real valued between 0 and 1) and recombined to give an estimated source $s(t)$.

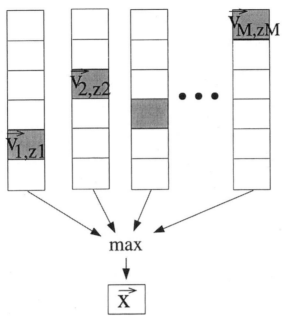

Figure 8.4. Generative model for the Factorial–MAX model. Each of M quantizers selects its z_m^{th} codebook vector v_{m,z_m} and these are combined using elementwise maximum to produce the final output x.

The MAXVQ model is useful in situations where there are multiple "objects", "sources" or "causes" in the world but there is some kind of occlusion or sparseness governing how the sources interact to produce observations. For example, as noted above, in clean speech recordings, the log spectrogram of a mixture of speakers is almost exactly the elementwise maximum of the log spectrograms of the individual speakers. For noisy mixtures of speech signals, each clean speaker and each noise source can be thought of as an independent cause contributing to the observed signal. We will use the short–time log power in linearly spaced narrow frequency bands as our vectors when analyzing speech with this model. (As another example, in range finding using laser or acoustic sensors, the distance reading in any direction is the minimum of the distances of all objects from the sensor in that direction.)

Formally, MAXVQ is a latent variable probabilistic model for D-dimensional data vectors x. The model consists of M vector quantizers, each with K_m codebook vectors v_m^k. Latent variables $z_m \in \{1, \ldots, K_m\}$ control which codebook vector each vector quantizer selects. Given these selections, the final output x is generated as a noisy version of the elementwise maximum of the selected codewords. If we assume that the each vector quantizer chooses its codebook entries independently with fixed rates π_m^k, then the model can be written as:

$$p(z_m = k|\pi) = \pi_m^k \qquad m \in \{1...M\}, k \in \{1...K_m\}$$

$$p(z) = \prod_m p(z_m), z = (z_1,...,z_M)$$

$$a_d = \arg\max_m (v_{md}^{z_m})$$

$$p(x_d|a_d, v, \Sigma) = N^+(x_d|v_{ad}^{z_a}, \Sigma_{ad})$$

$$p(x|v, \Sigma, \pi) = \sum_z p(x|\pi)p(x|z, v, \Sigma)$$

where z_m are latent variables, v_m^k are the codebook vectors, Σ_{md} are noise variances (shared across k), and M, K_m are structural size parameters chosen to control complexity. The distribution N_+ is the positive side of a Gaussian.

MAXVQ can be thought of as an exponentially large mixture of positive Gaussians having $(\Pi_m K_m)$ components, with the mean of each component constrained to be the elementwise max of some underlying parameters v_m^k. This technique, of representing an exponentially large codebook using a factorial expansion of a small number of underlying parameters has been very influential and successful in recent machine learning algorithms (Jojic and Frey, 2000, Ross and Zemel, 2003).

This model can also be extended through time to generate a Factorial–Max Hidden Markov Model (Roweis, 2001, Varga and Moore, 1990). There are some additional complexities, and the details of the heuristics used for inference are slightly different but in our experience the frame–independent MAXVQ model performs almost as well and so for simplicity, we will not discuss the full HMM model.

2.2.4   Parameter Estimation from Isolated Sources.

Given some isolated (clean) recordings of individual speech or noise sources, we can estimate the codebook means v_d^k, noise variances Σ_d and the selection probabilities π^k associated with the source's model by training a mixture density or a vector quantizer on the columns of a short–time narrowband log spectrogram. Some care must be taken in training to properly obey the nonnegativity assumption on the noise and to avoid too many codebook entries (mixture components) representing low energy (silent) segments (which are numerous in the data). Also, it is often advantageous to represent two or more adjacent (in time) columns of the spectrogram as a single input vector to allow the model to take some small advantage of temporal continuity.

2.2.5 Inference for Refiltering.

The key idea in this section is that the process of inference (i.e. deducing the values of the hidden variables given the parameters and observations) in a MAXVQ model performs a computation which is extremely useful for computing the masking signals required to perform refiltering for denoising or separation. Because the number of possible joint settings of the hidden selection variables z is exponentially large, we are usually only interested in finding the single most likely (MAP) setting of z given x or the N-best settings. (For unsupervised learning and likelihood computations we may also be interested in efficiently summing over all possible joint settings of z to compute the marginal likelihood of a given observation x.) Computing these Viterbi settings (or the sum) is intractable either by direct summation or by naive dynamic programming because of the factorial nature of the model. We must resort to branch-and-bound algorithms for efficient decoding or else approximations (e.g. variational methods) to estimate likely settings of z.

Once we have computed the MAP (or approximate) setting of z, we can use this to estimate the refiltering masking signals as follows: for each (overlapping) frame of the input spectrogram, set the masking signal to unity for every frequency at which the output proposed by the model corresponding to the source to be recovered is the maximum proposal over all models. Other frequencies have their masks set to zero. Actual refiltering is then performed by retaining the phase from the spectrogram of the original (noisy/mixed) recording, applying the (binary) masking signals to the log magnitude of each frequency, and reconstituting the clean signal using overlap-and-add reconstruction. The windowing function used to compute the original spectrogram must be known (or estimated) in order to remove its effect properly during refiltering.

2.2.6 Branch-and-Bound for Efficient Inference.

As discussed above, naive computation of the MAP joint settings of the hidden selection variables in MAXVQ is exponentially expensive. Fortunately, there is a clever branch and bound trick which can be used, based on the following observation: if $z_m = k$, we can upper bound the log likelihood we can achieve on a data case x, no matter what values the other $z_{m' \neq m}$ take on. The bound $\log p(x|z_m = k) \leq B_{mk}$ is constructed as follows (using constant Σ for simplicity):

$$B_{mk} = \frac{1}{2}\sum_d [x_d - v_{md}^k]_+^2 - \frac{D}{2}\log|\Sigma| - \log \pi_m^k \qquad (1.2)$$

where $[r]_+$ takes the max of zero and r. The intuition is that either v_m^k is *greater than* x along a certain dimension d of the output, in which case the error will be at least $(z_d - v_{md}^k)^2$ or else it is less than x along dimension d in which case the error on that dimension could potentially be zero.

This bound can be used to quickly search for the MAP setting of z given x as follows. For each $m \in \{1...M\}$ and each $k \in \{1...K_m\}$, compute the bound B_{mk}. Initially set the guess of the best configuration to the settings with the best bounds: $z_m^* = \arg\min_k B_{mk}$ and compute the true likelihood achieved by that guess: $l^* = \log p(x|z^*)$. Now, for each $m \in \{1...M\}$, we can eliminate all k for which $B_{mk} < l^*$. In other words, we can definitively say that certain codebook choices are impossible for certain models, independent of what other models choose because they would incur a minimum error worse that what has already been achieved. Now, for each m, and for all possible settings of k that remain for that m, systematically evaluate $\log p(x|z)$ and if it is less than l^*, eliminate the setting. If the likelihood is greater than l^*, we accept it as the new best setting and reset z^* and l^*, we also re-eliminate all settings of k that are now invalid because of this improved bound, and repeat until all settings have been either pruned or checked explicitly. This method is guaranteed to find the *exact* MAP setting, but it comes with no guarantees about its time complexity. In practice, however, we have found it to prune very aggressively and almost always find the MAP configuration in reasonable time. This technique is illustrated in figure 8.5.

2.3   Experiments with Factorial–MAX VQ

As an illustration of the methods presented above, we performed simple denoising and separation experiments using TIMIT prompts read by a single speaker and noise (babble) from the NOISEX database. Narrowband spectrograms we constructed from isolated, clean training examples of the speaker and noise. (Signals were downsampled to 12.5kHz, frames of length 512 were used with Hanning windows and frame shifts of 64 samples, resulting in one 257-vector of log energies each 5ms representing the signal over the last 40ms.) A simple vector–quantization codebook with 512 codewords was trained on the speech and one with 32 codewords was trained on the noise. Approximately 5 minutes of speech (with low energy frames eliminated) and 100 seconds of noise were used for training. A modified k-means algorithm which includes split-and-merge heuristics for finding good local optima was used. (We have also experimented with training \scaled" vector quantizers which cluster onto rays in the input space rather than on points, although this technique was not used in the results below.) The trained models were then used to perform MAXVQ inference on previously unseen test data, using the branch-and-bound technique. Based on this inference, refiltering was performed as described above to recover clean/isolated sources. In the denoising experiment, a 6 second speech segment was linearly mixed with 6 seconds of babble noise at 0dB SNR (equal power). Figure 8.6 shows the results of denoising with MAXVQ and also with a simple spectral subtraction trained on the same isolated noise sample as used for the VQ model. For the separation experiment, two different utterances, spoken by the same speaker, were mixed at equal power and the speech model was used (symmetrically) to perform MAXVQ inference. The results of this monaural separation are shown in figure 8.7. Of course, these results do not represent state of the art performance on either denoising or separation tasks, they are merely a proof of concept that the marriage of refiltering and inference in factorial models can be used for powerful speech processing tasks.

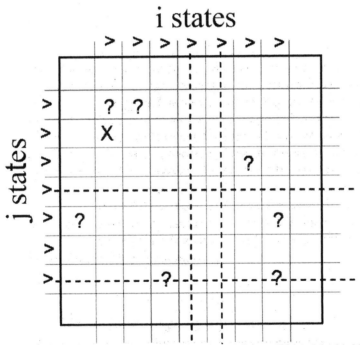

Figure 8.5. Branch and bound inference trick for a factorial–max VQ model with $M = 2$ quantizers. For each codebook selection that quantizer i could make, it is possible to compute a bound on the error (likelihood), regardless of the choice made by quantizer j. Similarly, for each codebook selection that j can make a bound (independent of i's choice) can be calculated. The current best joint selection of i and j is instantiated (shown by an x in the diagram), and its true error (likelihood) is computed. Any choices for either quantizer which are worse than this already achieved value are eliminated since they cannot possibly be involved in the MAP configuration. Remaining valid choices are explored, ordered by their bound values (indicated by ? in the diagram). Once all choices have either been explored or eliminated, we are guaranteed to have found the MAP configuration.

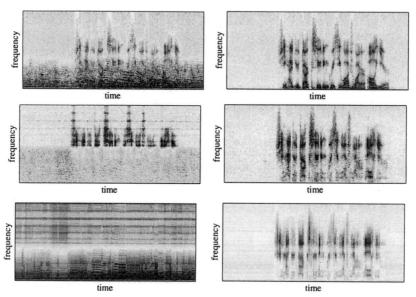

Figure 8.6. Denoising using MAXVQ. Top row: noisy input, original clean source. Middle row: spectral subtraction estimate (trained on isolated noise), MAXVQ estimate after exact branch-and-bound inference and refiltering (trained on isolated speech and noise). Bottom row: proposed codebook output sequence from noise model, proposed codebook output sequence from speech model.

2.4   MAXVQ: Discussion, Related and Future Work

We have argued that the *refiltering* approach to separation and denoising can be successfully achieved by using the inference step in a factorial model to provide the masking signals. Varga and Moore (Varga and Moore, 1990) proposed a factorial model for spectrograms (focusing on the factorial nature and using the log–max approximation) as did Gales and Young (Gales and Young, 1996) (focusing on the combination operation) but these models were used for speech recognition in the presence of noise only, and not for refiltering to do separation and denoising. In a series of papers, Green et.al. (Green *et al.*, 2001) have studied masking (refiltering) for denoising, but do not employ factorial model inference as an engine for finding masking signals. There have also been several approaches to monaural separation and denoising that operate mainly in the time domain, without using refiltering or factorial models. Cauwenberghs (Cauwenberghs, 1999) investigated separation based on maximizing periodic coherence, Wan and Nelson (Wan and Nelson, 1998) use nonlinear autoregressive networks and extended Kalman filtering.

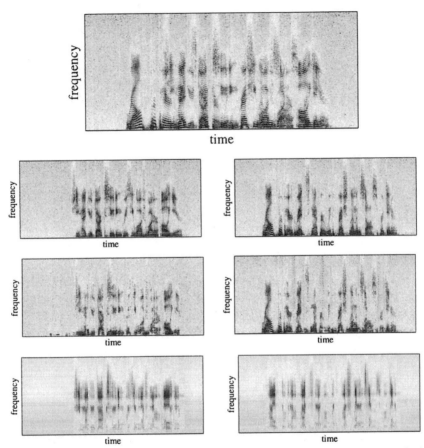

Figure 8.7. Monaural separation using MAXVQ. Top row: narrowband spectrogram of mixed input containing two different utterances spoken simultaneously by the same speaker. Second row: original isolated utterances. Third row: MAXVQ estimates of original utterances after exact branch-and-bound inference and refiltering (trained on isolated speech, not including this test example). Bottom row: proposed codebook output sequence for each stream.

Our work here and previously (Roweis, 2001) is closest in spirit to that of Ephraim et.al. (Ephraim *et al.*, 1989) who model speech using a HMM and noise using an AR model and then attempt to approximately infer the clean speech by alternating between Wiener filtering to find the noise and Viterbi decoding in the HMM. Logan and Moreno (Logan and Moreno, 1998) also investigated the use of factorial HMMs for modeling speech and found standard HMMs to be just as good, but they did not compose their model using the max of two underlying models, rather they learned separate parameters for each combination of states. Reyes et.al. (Reyes *et al.*, 2003) investigated factorial HMMs for separation but using multi-channel inputs.

The main challenge for future work is to develop techniques for learning from only mixed/noisy data, without requiring clean, isolated examples of individual sources or noises at training time. In a maximum likelihood formulation of this purely unsupervised learning setup, we would be given many realizations of x from the model, $X = [x^1, x^2,..., x^N]$, assumed to be IID, and we attempt to adjust the model parameters so as to make the observed data more likely. Using the view proposed above, in which MAXVQ is seen as a very large mixture of Gaussians with parameter tying on the means, we can learn the parameters of a MAXVQ model using the standard EM algorithm for maximum likelihood. However, this requires summing over all K^M possible settings of z explicitly. If K^M is small enough for this to be feasible, then this is one possible way to do learning. Otherwise, approximate inference techniques must be used to allow tractable computations.

Along this line, we are investigating a technique which gives approximate rather than exact answers but has a fixed and known time complexity. This idea is to introduce a *factorized variational distribution* which tries to approximate the true (joint) posterior as well as possible. In this setup, we approximate the true posterior $p(z|x)$ with a factorized posterior $q(z|x)=\Pi_m q(z_m|x)$ and proceed to find the functions $q(z_m|x)$ which maximize a lower bound on the data likelihood.

3   A SEGMENTAL HMM FOR SPEECH WAVEFORMS

In the following section, we turn our attention to another simple generative model, this time one which operates directly on the speech waveform. Pursuing inference in this model leads to a purely time domain approach to pitch processing which identifies waveform samples at the boundaries between glottal pulse periods (in voiced speech) or at the boundaries between unvoiced segments. An efficient algorithm for inferring these boundaries is derived from a simple probabilistic generative model for segments, which gives excellent results on pitch tracking, voiced/unvoiced detection and timescale modification.

3.1   Speech Segments in the Time Domain

Processing of speech signals directly in the time domain is commonly regarded to be difficult and unstable, due to fact that perceptually very similar utterances exhibit very large variability in their raw waveforms. As a result, by far the most common preprocessing step for most speech systems is to convert the raw waveform into a time–frequency representation, using a variety of spectral analysis and filterbank techniques. In this section we explore a purely time domain approach to speech processing in which we identify the samples at the boundaries between glottal pulse periods (in voiced speech) or at the boundaries between unvoiced segments of similar spectral shape.

Having identified these segment boundaries, we can perform a variety of important low level speech analysis and manipulation operations directly and conveniently. For example, we make a voiced/unvoiced decision on each segment by examining the periodicity of the waveform in that segment only. In voiced segments we can estimate the pitch as the reciprocal of the segment length. Timescale modification without pitch or format distortion can be achieved by stochastically eliminating or replicating segments in the time domain directly. More sophisticated operations, such as pitch modification, gender and voice conversion, and companding (volume equalization) are also naturally performed by operating on waveform segments one by one without the need for a cepstral or other such representation. In effect, our model chops up the original speech wave into natural "atomic" units which can be examined or manipulated in very flexible ways.

The computational challenge with this approach is in efficiently and robustly identifying the segment boundaries, across silence, unvoiced and voiced segments. In this section we describe a segmental Hidden Markov Model (Achan *et al.*, 2004), defined on variable length sections of the time domain waveform, and show that performing inference in this model allows us to identify segment boundaries and achieve excellent results on the speech processing tasks described above.

3.2   A probabilistic generative model of time–domain speech segments

The goal of our algorithm is to break the time domain speech signal $s^1,...,s^N$ into a set of segments, each of which corresponds to a glottal pulse period or a segment of unvoiced colored noise. Let b_k denote the time index of the beginning of the kth segment and $s_k=(s_{bk},...,s_{bk+1}-1)$ denote the waveform in the kth segment, where $k=1,...,K$ indexes segments. Our algorithm searches for the segment boundaries, $b_1,b_2,...,b_{k+1}$, so that each segment can be accurately modeled as a time–warped, amplitude–scaled and amplitude–shifted version of the previous segment. We denote the transformation used to map segment s_{k-1} into segment s_k by T_k.

Given the segment boundaries $b_1,...,b_{k+1}$ and the transformations $T_1,...,T_k$ we assume the probability of each segment depends only on the previous segment and the transformation for that segment: in other words we assume the segments are generated by first order Markov chain:

$$P(s_1, s_2, \ldots, s_K | b_1, \ldots, b_{K+1}, T_1, \ldots, T_K)$$

$$= \prod_{k=1}^{K} P(s_k | s_{k-1}, b_{k-1}, b_k, b_{k+1}, T_k) \qquad (1.3)$$

Each segment is modeled as a noisy copy of the transformed version of the previous segment. These assumptions simplify the inference and estimation algorithm described below. Of course, number of segments and the segment boundaries are unknown and must be inferred from the speech wave: this inference is the main computation performed by our algorithm.

For concreteness, we assume that each successive segment s_k is equal to a transformed version of the previous segment, plus isotropic, zero–mean normal noise with variance σ_k^2. Denoting the transformed version of segment $k{-}1$ by $T_k s_{k-1}$, the conditional probability density of s_k is:

$$P(s_k | s_{k-1} b_{k-1} b_k b_{k+1} T_k) = \frac{1}{2\pi\sigma_k^{2^{b_{k+1} - b_k + 1/2}}}$$

$$\bullet \exp\left(-\frac{1}{2\sigma_k^2}(s_k - T_k s_{k-1})^T (s_k - T_k s_{k-1})\right) \qquad (1.4)$$

The noise levels $\sigma_2^2, \ldots, \sigma_K^2$ are estimated automatically by the inference procedure along with the segment boundaries (As the boundary condition of the Markov chain, we assume that the segment before the first is a vector of all zeros ($s_0 = 0$) and hence the probability density of the initial segment is given by $(2\pi\sigma_1^2)^{-b_2/2} \exp(-s_1^T s_1 / 2\sigma_1^2)$. We also set σ_1^2 equal to the variance of all time–domain samples, since *a priori* we do not know what the content of the initial segment should be.)

We assume that the boundaries and transformations are independent, and that the prior distribution over transformations is uniform on some bounded set. In our experiments, we parameterize the transformation by $T_k(\alpha_k, \beta_k, \gamma_k)$, where α_k, β_k and γ_k are time–warp, amplitude–scaling and amplitude–shift. We use a prior that is uniform over a 3-dimensional hypercube that includes all reasonable values for these parameters.

Generally the joint prior probability mass function on segment boundaries $P(b_1, \ldots, b_k)$ can be quite complex. Since the computational complexity of the inference algorithm will depend on the number of allowed configurations of segment boundaries, we use a prior that is nonzero only on an appropriate subset of configurations. In particular, we exploit a very simple heuristic (first suggested by John Hopfield in 1998) by *restricting segments to begin and end only on zero crossings of the signal* (or possibly only on upward or downward going zero crossings). This restriction also allows arbitrary segments to be relocated beside each other and still preserve waveform continuity, which will be important in our later applications. To further restrict the range of inferred segment lengths, we require that $\Delta_{min} \le b_k - b_{k-1} \le \Delta_{max}$, where Δ_{min} and Δ_{max} are the minimum and maximum segment lengths, satisfying $\Delta_{max} > \Delta_{min} > 0$. These minimum and maximum segment lengths are chosen to represent the widest possible range of pitch periods we expect to see in our signals. We assume the probability $P(b_1,...,b_{k-1})$ is otherwise uniform, subject to the above constraints. The number of segments K is also unknown, and its optimal value is inferred automatically as well. We assume that $b_1 = 1$ (the first segment begins on the first signal sample) and that $b_{K=1} = N+1$ (the last segment ends on the last signal sample).

The joint distribution over segments, segment boundaries and transformations can now be written as:

$$P(s_1,...,s_K,b_1,...,b_{K+1},T_1,...,T_K) \propto P(b_1,...,b_{K+1})$$

$$\prod_{k-1}^{K} P(T_k)P(s_k|s_{k-1},b_{k-1},b_k,b_{k+1},T_k) \qquad (1.5)$$

where $P(b_1,...,b_{k+1})$ enforces the constraints on the boundaries, constraints on the allowable limits of the time domain scale, amplitude–domain scale and amplitude–domain shift are enforced by $(\prod_{k=1}^{K} P(T_k))$ although these constraints rarely affect the optimization.

3.3   Using dynamic programming to efficiently infer segment boundaries and transformations

Given a time–domain signal, the computational task now at hand is to determine the segment boundaries and transformations. Of course, the number of valid configurations of the boundary variables is exponential in the length of the waveform, so exhaustive search would be intractable. Fortunately, the optimal segmentation (which maximizes likelihood, or equivalently, minimizes the mismatch penalties) can be found using a generalized dynamic programming algorithm.

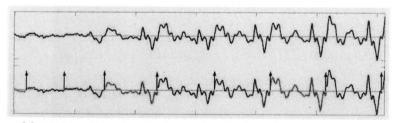

Figure 8.8. (*top*) Input signal, notice the transition from unvoiced to voiced region. (*bottom*) Inferred maximum likelihood segmentation found using generalized dynamic programming. The upward arrows are used to mark the inferred segment boundaries.

First, note that according to 1.5, given the boundary variables, the MAP estimates of the transformations can be computed locally:

$$\underset{T_k}{\arg\max}\, P(s_1,\ldots,s_K,b_1,\ldots,b_{K+1},T_1,\ldots,T_K)$$

$$= \underset{T_k}{\arg\max}\, P(T_k)P(s_k|s_{k-1},b_{k-1},b_k,b_{k+1},T_k) \qquad (1.6)$$

In particular, the time–warping is unique and is given by $\alpha_k=(b_{k+1}-b_k)/(b_k-b_{k-1})$. The warped version of s_{k-1} is denoted by $\hat{s}_{k-1}$ and can be obtained using standard signal processing techniques for time–domain interpolation or decimation. Note that whereas s_{k-1} contains b_k-b_{k-1} samples, $\hat{s}_{k-1}$ contains $b_{k+1}-b_k$ samples. The amplitude–domain scale β_k and shift γ_k are obtained by performing a least–squares regression of $\hat{s}_{k-1}$ onto s_k, i.e. by solving

$$\underset{\beta_k,\gamma_k}{\arg\min}\ \sum_{j=1}^{b_{k+1}-b_k} \left(\beta_k\hat{s}_{k-1}(j) + \gamma_k - s_k(j)\right)^2, \qquad (1.7)$$

where (j) indexes the elements of s_k and $\hat{s}_{k-1}$. After optimizing β_k and γ_k, the estimate of the variance σ_k^2 is set to the argument in the above minimization, divided by $b_{k+1}-b_k$. For a given configuration of b_{k-1},b_k,b_{k+1}, we denote the optimal transformation obtained in the above fashion by T_k^*.

Thus, the search is one over boundary segment positions, which for efficiency we constrain to lie only at (just after) zero crossings of the waveform. Finding the optimal segmentation requires performing dynamic programming, using a table indexed with two adjacent boundary points. In order to make the optimization Markovian, we must actually consider *adjacent pairs* of boundary points (b_{k-1},b_k) as the states in the dynamic programming. In particular, we fill in a table C whose entry $C(m,n)$ holds the best possible log likelihood of the segmentation ending with the segment defined by the mth zero crossing at its left edge and the nth zero crossing at its right edge. We can iteratively fill in this table forwards for all values $m<n$, by using the following recursion:

$$C(m, n) = \underset{i}{\arg\max}[C(m, i) + \log P(s_k|s_{k-1},b_{k-1} = m,b_k = i,b_{k+1} = n,T_k)] \qquad (1.8)$$

3.4   Segmental HMM Experiments

We have applied our segmental inference procedure to clean, wideband recordings of single–talker speech, from both males and females taken from the Keele pitch reference dataset (Plante *et al.*, 1995) and from the Wall Street Journal (WSJ) corpus. Dynamic programming was applied with segment length thresholds of Δ_{min}=3ms and Δ_{max}=25ms (corresponding to pitch range of 40Hz-333Hz) to find the optimal segmentation of the raw waveforms directly.

We can apply the results of our segment inference algorithm to a wide range of speech processing tasks, as discussed below. By replicating or deleting some or all of the inferred segments, we can easily achieve high quality timescale modification without changing the perceived pitch or formant structure of the utterance. By examining the periodicity of each segment, we can attempt to distinguish voiced from unvoiced portions of the waveform. In voiced regions, we can directly estimate the pitch by taking the reciprocal of the segment length. Below, we present results on timescale modification, voiced/unvoiced discrimination, and pitch tracking. Other applications such as gender and voice conversion, companding and concert hall effects are also possible. We emphasize that all the experiments were performed in *time domain* using the inferred pitch periods.

3.4.1   Voicing Detection and Pitch Tracking.

For voicing detection and pitch tracking, we evaluated the estimates obtained using our algorithm using the Keele dataset, since it has ground truth values for these quantities. (In particular, the Keele data has utterances spoken by both male and female speakers and includes a reference estimate for the fundamental frequency at a resolution of 10ms. Each utterance is approximately 30 seconds long and the sampling frequency is 20kHz.)

Once the waveform segments are inferred by the algorithm, we can estimate the periodicity of each segment in a simple way by computing the discrete Fourier transform of the segment waveform and then reconstructing it using a limited number of Fourier coefficients. (This is illustrated in figure 8.9.)

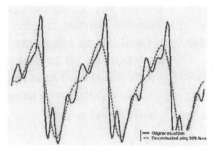

Figure 8.9. Simple voicing detection given waveform segmentation. Each segment is reconstructed using a small number of Fourier coefficients. Segments whose reconstruction error is below some threshold (and whose energy is above the silence threshold) are tagged as voiced. Examples above show typical voiced (left) and unvoiced (right) segments and their reconstructions.

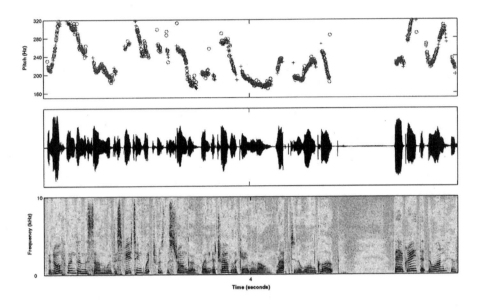

Figure 8.10. (*top*) Pitch estimates using segmental HMM for a female speaker in the Keele dataset. Notice that the inferred pitch (red circle) consistently agrees with the reference provided (blue plus mark). Further, our approach clearly discriminates between voiced/ unvoiced regions (samples without reference estimates are unvoiced). (*center*) input time domain signal (*bottom*) spectrogram of input.

Since unvoiced regions tend to be much less periodic, they will have a substantially larger reconstruction error than voiced regions and by selecting an appropriate threshold, we can discriminate between voiced and unvoiced segments. Our method was able to correctly identify 87.2% of the voiced segments averaged over all the 10 utterances of males and females in the Keele dataset. In Fig. 8.10, the true unvoiced regions are the segments without any reference pitch shown, and the unvoiced regions detected by our algorithm are those without estimated pitches.

Pitch tracking is trivially achieved by taking the reciprocal of the segment lengths in the voiced regions. Results for a single utterance in the Keele dataset spoken by a female speaker is shown in Fig. 8.10. Pitch estimates obtained using our approach are very consistent with the reference estimates, similar performance was obtained on other utterances in the dataset as well. Averaged over 10 utterances the median absolute pitch error was 9Hz.

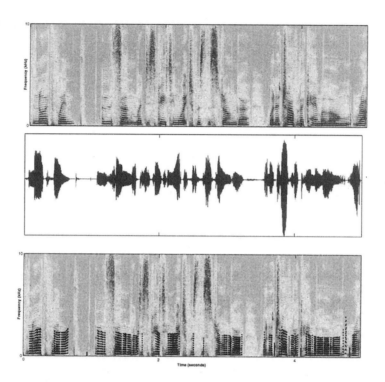

Figure 8.11. (*Middle and Top*) Time domain signal and the corresponding spectrogram (*Bottom*) The spectrogram of the signal is marked with the pitch estimates obtained using our algorithm (blue marker), for clarity we have marked only the first 10 integer multiples of the fundamental frequency.

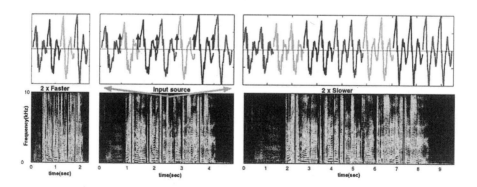

Figure 8.12. The spectrogram of time scale modified faster and slower versions of a signal are shown. The actual time domain operation is shown on top for a particular time instant in the spectrogram.

It is well known that excitation for voiced speech manifests as sharp bursts at integer multiples of fundamental frequency. In Fig. 8.11, we have shown a few integer multiples of the fundamental frequency of a signal on its spectrogram using pitch estimates obtained from the application of our algorithm.

3.4.2   Time Scale Modification.

For timescale modification experiments, we have used utterances from the WSJ corpus. Once the segments are identified by our algorithm, we can play the signal twice as fast by deleting every other segment and concatenating the remaining ones, similarly by replicating each segment we can achieve the effect of playing the at half the speed (two times slower), this is further illustrated in Fig. 8.12. This approach is substantially different from methods such as (Roucos and Wilgus, 1985) that manipulate spectrograms. By doing all of our operations directly in the time domain we never need to worry about inconsistent phase estimates.

3.5   Segmental HMM: Discussion and Conclusions

We have presented a simple segmental Hidden Markov Model for generating a speech waveform and derived an efficient algorithm for approximate inference in the model. Applied to an observed signal, this inference algorithm operates entirely in the time domain and is capable of identifying the boundaries of glottal pulse periods in voiced speech and of unvoiced segments. Using these inferred boundaries we are able to easily and accurately detect voicing, track pitch and modify the timescales. We are investigating other possible applications of the same basic model, including voice conversion, volume equalization and reverberant filtering.

4   CONSTRAINED HIDDEN MARKOV MODELS FOR ARTICULATORY MODELING

Structured time–series are generated by systems whose underlying state variables change in a continuous way but whose state to output mappings are highly nonlinear, many to one and not smooth. Probabilistic unsupervised learning for such sequences requires models with two essential features: latent (hidden) variables and *topology* in those variables.

By thinking of each state in a hidden Markov model as corresponding to some spatial region of a fictitious *topology space* it is possible to naturally define neighbouring states of any state as those which are connected in that space. The transition matrix of the HMM can then be constrained to allow

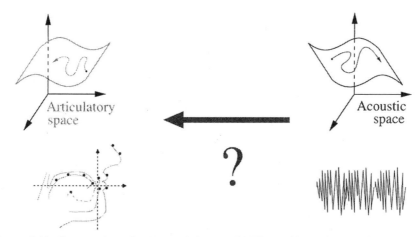

Figure 8.13. Can you hear the shape of the mouth? The problem of recovering articulator motions from acoustics is a classic inversion problem involving physics, speech science and statistical signal processing.

transitions only between neighbours, this means that all valid state sequences correspond to connected paths in the topology space. This strong constraint makes structure discovery in sequences easier. We show how such *constrained HMMs* can learn to discover underlying structure in complex sequences of high dimensional data, and apply them to the problem of recovering mouth movements from acoustics in continuous speech. This problem has a long history in speech science, and exemplifies exactly the sort of structured time series analysis problem discussed above.

4.1　Constrained HMMs as latent variable models

Hidden Markov models (HMMs) can be thought of as dynamic generalizations of discrete state static data models such as Gaussian mixtures, or as discrete state versions of linear dynamical systems (LDSs) (which are themselves dynamic generalizations of continuous latent variable models such as factor analysis). While both HMMs and LDSs provide probabilistic latent variable models for time–series, both have important limitations. Traditional HMMs have a very powerful model of the relationship between the underlying state and the associated observations because each state stores a private distribution over the output variables. This means that any change in the hidden state can cause arbitrarily complex changes in the output distribution. However, it is extremely difficult to capture reasonable dynamics on the discrete latent variable because in principle any state is reachable from any other state at any time step and the next state depends only on the current state. This allows only "random jump" movement in the hidden state space. For example, one well known difficulty is that the lifetime of any single state is distributed according to a decaying exponential, which is often an inappropriate distribution for state dwell times. LDSs, on the other hand, have an extremely impoverished representation of the outputs as a function of the latent variables since this transformation is restricted to be global and linear: a single matrix captures the state to output mapping and it is applied uniformly regardless of location in the state space. But it is somewhat easier to capture state dynamics since the state is a multidimensional vector of continuous variables on which a matrix "ow" is acting, this enforces some continuity of the latent variables across time. *Constrained hidden Markov models* (Roweis, 2000) address the modeling of state dynamics by building some topology into the hidden state representation. The essential idea is to constrain the transition parameters of a conventional HMM so that the discrete–valued hidden state evolves in a structured way.[6] In particular, below we consider parameter restrictions which constrain the state to evolve as a discretized version of a continuous multivariate variable, i.e. so that it inscribes only connected paths in some space. This lends a physical interpretation to the discrete state trajectories in an HMM.

4.2   An illustrative game

Consider playing the following game: divide a sheet of paper into several contiguous, non-overlapping regions which between them cover it entirely. In each region inscribe a symbol, allowing symbols to be *repeated* in different regions. Place a pencil on the sheet and move it around, reading out (in order) the symbols in the regions through which it passes. Add some *noise* to the observation process so that some fraction of the time incorrect symbols are reported in the list instead of the correct ones. The game is to reconstruct the configuration of regions on the sheet from only such an ordered list(s) of noisy symbols. Of course, the absolute scale, rotation and reflection of the sheet can never be recovered, but learning the essential *topology* may be possible.[6] Figure 8.14 illustrates this setup.

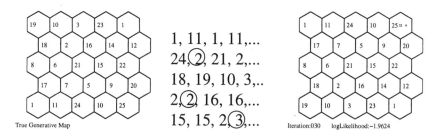

Figure 8.14. (left) True map which generates symbol sequences by random movement between connected cells. (centre) An example noisy output sequence with noisy symbols circled. (right) Learned map after training on 3 sequences (with 15% noise probability) each 200 symbols long. Each cell actually contains an entire distribution over all observed symbols, though in this case only the upper right cell has significant probability mass on more than one symbol. Only the top three symbols of each cell's histogram are show, with *font size proportional to the square root of probability* (to make ink roughly proportional).

Without noise or repeated symbols, the game is easy (non-probabilistic methods can solve it) but in their presence it is not. One way of mitigating the noise problem is to do statistical averaging. For example, one could attempt to use the average separation *in time* of each pair of symbols to define a dissimilarity between them. It then would be possible to use methods like multi-dimensional scaling or a sort of *Kohonen mapping though time*[7] to explicitly construct a configuration of points obeying those distance relations. However, such methods still cannot deal with many-to-one state to output mappings (repeated numbers in the sheet) because by their nature they assign a unique spatial location to each symbol.

Playing this game is analogous to doing unsupervised learning on structured sequences. (The game can also be played with continuous outputs, although often high–dimensional data can be effectively clustered around a manageable number of prototypes, thus a vector timeseries can be converted into a sequence of symbols.) Constrained HMMs incorporate latent variables with topology yet retain powerful nonlinear output mappings and can deal with the difficulties of noise and many–to–one mappings mentioned above, so they can "win" our game (see fig. 8.1). The key insight is that the game generates sequences exactly according to a hidden Markov process whose transition matrix allows only transitions between neighbouring cells and whose output distributions have most of their probability on a single symbol with a small amount on all other symbols to account for noise.

4.3   Model definition: state topologies from cell packings

Defining a constrained HMM involves identifying each state of the underlying (hidden) Markov chain with a spatial cell in a fictitious *topology space*. This requires selecting a *dimensionality d* for the topology space and choosing a *packing* (such as hexagonal or cubic) which fills the space. The number of cells in the packing is equal to the number of states M in the original Markov model. Cells are taken to be all of equal size and (since the scale of the topology space is completely arbitrary) of unit volume. Thus, the packing covers a volume M in topology space with a side length l of roughly $l=M^{1/d}$. The dimensionality and packing together define a vector–valued function $x(m)$, $m=1,...,M$ which gives the location of cell m in the packing. (For example, a cubic packing of d dimensional space defines $x(m+1)$ to be $[m,m/l,m/l^2,...,m/l^{d-1}]\mathrm{mod}\,l$.) (here a mod b denotes the remainder after dividing a by b). State m in the Markov model is assigned to cell m in the packing, thus giving it a location $x(m)$ in the topology space. Finally, we must choose a *neighbourhood rule* in the topology space which defines the neighbours of cell m, for example, all "connected" cells, all face neighbours, or all those within a certain radius. (For cubic packings, there are 3^d-1 connected neighbours and $2d$ face neighbours in a d dimensional topology space.) The neighbourhood rule also defines the boundary conditions of the space – e.g. periodic boundary conditions would make cells on opposite extreme faces of the space neighbours with each other.

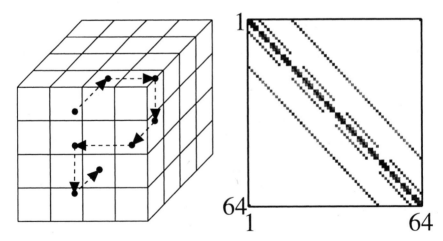

Figure 8.15. (left) Physical depiction of the topology space for a constrained HMM with $d=3$, $l=4$ and $M=64$ showing an example state trajectory. (right) Corresponding transition matrix structure for the 64-state HMM computed using face–centred cubic packing. The gaps in the inner bands are due to edge effects.

The transition matrix of the HMM is now *preprogrammed* to only allow transitions between neighbours. All other transition probabilities are set to zero, making the transition matrix very sparse. (We set all permitted transitions to be equally likely.) Now, *all valid state sequences in the underlying Markov model represent connected ("city block") paths through the topology space.* Figure 8.15 illustrates this for a three-dimensional model.

4.4   State inference and learning

The constrained HMM has exactly the same inference procedures as a regular HMM: the *forward–backward algorithm* for computing state occupation probabilities and the *Viterbi decoder* for finding the single best state sequence. Once these discrete state inferences have been performed, they can be transformed using the state position function $x(m)$ to yield probability distributions over the topology space (in the case of forward–backward) or paths through the topology space (in the case of Viterbi decoding). This transformation makes the outputs of state decodings in constrained HMMs comparable to the outputs of inference procedures for continuous state dynamical systems such as Kalman smoothing.

The learning procedure for constrained HMMs is also almost identical to that for HMMs. In particular, the EM algorithm (Baum-Welch) is used to update model parameters. The crucial difference is that the transition probabilities which are precomputed by the topology and packing are never updated during learning. In fact, having a preprogrammed and fixed transition matrix makes learning much easier in some cases. Not only do the transition probabilities not have to be learned, but their structure constrains the hidden state sequences in such a way as to make the learning of the output parameters much more efficient when the underlying data really does come from a spatially structured generative model. Notice that in this case, each part of state space had only a single output (except for noise) so the final learned output distributions became essentially minimum entropy. But constrained HMMs can in principle model stochastic or multimodal output processes since each state stores an entire private distribution over outputs.

4.5   Recovery of mouth movements from speech audio

We have applied the constrained HMM approach described above to the problem of recovering mouth movements from the acoustic waveform in human speech. Data containing simultaneous audio and articulator movement information was obtained from the University of Wisconsin X-ray microbeam database (Westbury, 1994). Eight separate points (four on the tongue, one on each lip and two on the jaw) located in the midsaggital plane of the speaker's

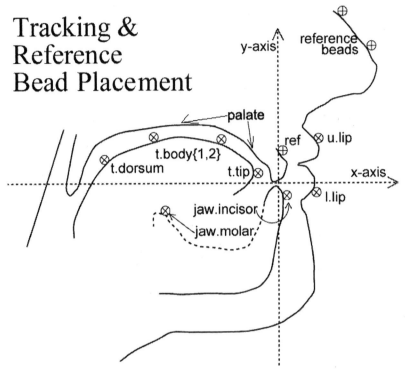

Figure 8.16. Midsaggital locations of tracking beads in the University of Wisconsin X-ray microbeam articulatory dataset.

head were tracked while subjects read various words, sentences, paragraphs and lists of numbers. The x and y coordinates (to within about ±1 mm) of each point were sampled at 146Hz by an X-ray system which located gold beads attached to the feature points on the mouth, producing a 16-dimensional vector every 6.9ms. The audio was sampled at 22kHz with roughly 14 bits of amplitude resolution but in the presence of machine noise.

How do these data relate to the general task introduced above? These data are well suited to the constrained HMM architecture. They come from a system whose state variables are known, because of physical constraints, to move in connected paths in a low degree-of-freedom space. In other words the (normally hidden) articulators (movable structures of the mouth), whose positions represent the underlying state of the speech production system,[8] move slowly and smoothly. The observed speech signal – the system's output – can be characterized by a sequence of short–time spectral feature vectors, often known as a *spectrogram*. In the experiments reported here, we have characterized the audio signal using 12 line spectral frequencies (LSFs) measured every 6.9ms (to coincide with the articulatory sampling rate) over a 25ms window. These LSF vectors characterize only the *spectral shape* of the speech waveform over a short time but not its energy. Average energy (also over a 25ms window every 6.9ms) was measured as a separate one dimensional signal. Unlike the movements of the articulators, the audio spectrum/energy can exhibit quite abrupt changes, indicating that the mapping between articulator positions and spectral shape is not smooth. (Compare the sampling rate of 22kHz for the acoustic signal with 146Hz for the articulators.) Furthermore, the mapping is many to one: *different* articulator configurations can produce very similar spectra (see below).

The unsupervised learning task, then, is to explain the complicated sequences of observed spectral features (LSFs) and energies as the outputs of a system with a low-dimensional state vector that changes slowly and smoothly. In other words, can we learn the parameters[9] of a constrained HMM such that connected paths through the topology space (state space) generate the acoustic training data with high likelihood? Once this unsupervised learning task has been performed, we can (as shown below) relate the learned trajectories in the topology space to the true (measured) articulator movements.

While many models of the speech production process predict the many-to-one and non-smooth properties of the articulatory to acoustic mapping, it is useful to confirm these features by looking at real data. Figure 8.17 shows the experimentally observed distribution of articulator configurations used to produce similar sounds. It was computed as follows. All the acoustic and articulatory data for a single speaker are collected together. Starting with some sample called the *key sample*, we find the 1000 samples "nearest" to this key by two measures: articulatory distance, defined using the Mahalanobis norm between two position vectors under the global covariance of all positions for the appropriate speaker, and spectral shape distance, again defined using the Mahalanobis norm but now between two line spectral frequency vectors using the global LSF covariance of the speaker's audio data. In other words, I find the 1000 samples that "look most like" the key sample in mouth shape and that "sound most like" the key sample in spectral shape. we then plot the tongue bead positions of the key sample (as a thick cross), and the 1000 nearest samples by mouth shape (as a thick ellipse) and spectral shape (as dots). The points of primary interest are the dots, they show the distribution of tongue positions used to generate very similar sounds. (The thick ellipses are shown only as a control to ensure that many nearby points to the key sample *do* exist in the dataset.) Spread or multimodality in the dots indicates that many *different* articulatory configurations are used to generate the *same* sound.

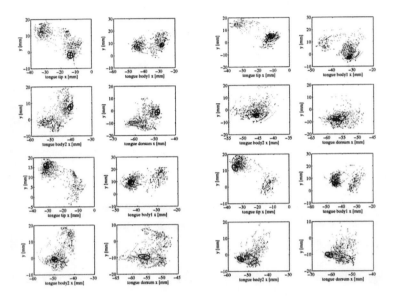

Figure 8.17. Inverse mapping from acoustics to articulation is ill–posed in real speech production data. Each group of four articulator–space plots shows the 1000 samples in the entire dataset which are "nearest" to one key sample (thick cross). The dots are the 1000 nearest samples using an acoustic measure based on line spectral frequencies. Spread or multimodality in the dots indicates that many different articulatory configurations are used to generate very similar sounds. Only the positions of the four tongue beads have been plotted. Four examples (with different key samples) are shown, one each group of four panels. The thick ellipses (shown as a control) are the two–standard deviation contour of the 1000 nearest samples using an articulatory position distance metric.

Why not do direct supervised learning from short–time spectral features (LSFs) to the articulator positions? The ill–posed nature of the inverse problem as shown in figure 8.17 makes this impossible. To illustrate this difficulty, we have attempted to recover the articulator positions from the acoustic feature vectors using Kalman smoothing on a LDS. In this case, since we have access to both the hidden states (articulator positions) and the system outputs (LSFs) we can compute the optimal parameters of the model directly. (In particular, the state transition matrix is obtained by regression from articulator positions and velocities at time t onto positions at time $t+1$, the output matrix by regression from articulator positions and velocities onto LSF vectors, and the noise covariances from the residuals of these regressions.) Figure 8.18b shows the results of such smoothing, the recovery is quite poor, even when the test utterance is included in the training set used to estimate model parameters.

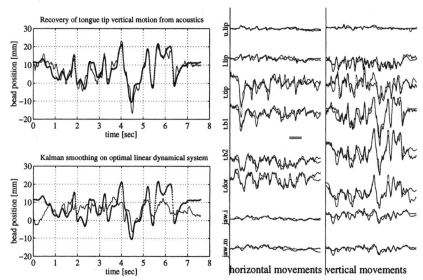

Figure 8.18. (A) Recovered articulator movements using state inference on a constrained HMM. A four-dimensional model with 4096 states (d=3,l=6,M=216) was trained on data (all beads) from a single speaker but not including the test utterance shown. Dots show the actual measured articulator movements for a single bead coordinate versus time, the thin lines are estimated movements from the corresponding acoustics. (B) Unsuccessful recovery of articulator movements using Kalman smoothing on a global LDS model. All the (speaker–dependent) parameters of the underlying linear dynamical system are known, they have been set to their optimal values using the true movement information from the training data. Furthermore, for this example, the test utterance shown was included in the training data used to estimate model parameters. (C) All 16 bead coordinates, all vertical axes are the same scale. Bead names are shown on the left. Horizontal movements are plotted in the left–hand column and vertical movements in the right–hand column. The separation between the two horizontal lines near the centre of the right panel indicates the machine measurement error.

Constrained HMMs can be applied to this recovery problem, as previously reported (Roweis and Alwan, 1997). (Earlier results used a small subset of the same database that was not continuous speech and did not provide the hard experimental verification (fig. 8.17) of the many-to-one problem.)

The basic idea is to train (unsupervised) on sequences of acoustic–spectral features and then map the topology space state trajectories onto the measured articulatory movements. Figure 8.18 shows movement recovery using state inference in a four-dimensional model with 4096 states ($d=4,l=8,M=4096$) trained on data (all beads) from a single speaker. (Naive unsupervised learning runs into severe local minima problems. To avoid these, in the simulations shown above, models were trained by slowly annealing two learning parameters[10]: a term ε^{β} was used in place of the zeros in the sparse transition matrix, and γ_t^β was used in place of $\gamma_t = p(m_t|observations)$ during inference of state occupation probabilities. Inverse temperature β was raised from 0 to 1.) To infer a continuous state trajectory from an utterance after learning, we first do Viterbi decoding on the acoustics to generate a discrete state sequence m_t and then interpolate smoothly between the positions $x(m_t)$ of each state.

After unsupervised learning, a single linear fit is performed between these continuous state trajectories and actual articulator movements on the training data. (The model cannot discover the units system or axes used to represent the articulatory data.) To recover articulator movements from a previously unseen test utterance, we infer a continuous state trajectory as above and then apply the single linear mapping (learned only once from the training data).

4.6   Constrained HMMs for Articulatory Data: Conclusions, extensions and related work

By enforcing a simple constraint on the transition parameters of a standard HMM, a link can be forged between discrete state dynamics and the motion of a real–valued state vector in a continuous space. For complex time–series generated by systems whose underlying latent variables do in fact change slowly and smoothly, such constrained HMMs provide a powerful unsupervised learning paradigm. They can model state to output mappings that are highly nonlinear, many to one and not smooth. Furthermore, they rely only on well understood learning and inference procedures that come with convergence guarantees.

Results on synthetic and real data show that these models can successfully capture the low-dimensional structure present in complex vector time–series. In particular, we have shown that a speaker dependent constrained HMM can accurately recover articulator movements from continuous speech to within the measurement error of the data. This acoustic to articulatory inversion problem has a long history in speech processing (see e.g. Schroeter and Sondhi, 1994, and references therein). Many previous approaches have attempted to exploit the smoothness of articulatory movements for inversion or modeling: Hogden et.al (e.g. Nix and Hogden, 1999) provided early inspiration for these ideas, but do not address the many-to-one problem, Simon Blackburn (Blackburn and Young, 1996) has investigated a *forward* mapping from articulation to acoustics but does not explicitly attempt inversion, early work at Waterloo (Ramsay and Deng, 1994) suggested similar constraints for improving speech recognition systems but did look at real articulatory data, more recent work at Rutgers (Chennoukh *et al.*, 1997) developed a very similar system much further with good success. Perpiñán (Carreira-Perpiñán, 2000), considers a related problem in sequence learning using EPG speech data as an example.

While in this section we have described only "diffusion" type dynamics (transitions to all neighbours are equally likely) it is also possible to consider *directed* flows which give certain neighbours of a state lower (or zero) probability. The left-to-right HMMs mentioned earlier are an example of this for one-dimensional topologies. For higher dimensions, flows can be derived from discretization of matrix (linear) dynamics or from other physical/structural constraints. It is also possible to have many connected local ow regimes (either diffusive or directed) rather than one global regime as discussed above, this gives rise to *mixtures* of constrained HMMs which have block–structured rather than banded transition matrices. Smyth (Smyth, 1997) has considered such models in the case of one-dimensional topologies and directed flows, we have applied these to learning character sequences from English text. Another application I have investigated is map learning from multiple sensor readings. An explorer (robot) navigates in an unknown environment and records at each time many local measurements such as altitude, pressure, temperature, humidity, etc. We wish to reconstruct from only these sequences of readings the topographic maps (in each sensor variable) of the area as well as the trajectory of the explorer. A final application is tracking (inferring movements) of articulated bodies using video measurements of feature positions.

5   SUMMARY

In this chapter, we have explored the use of inference in probabilistic generative models as a powerful signal processing tool for speech and audio. The basic paradigm explored was to design a simple model for the data we observe in which the key quantities that we would eventually like to compute appear as hidden (latent) variables. By executing probabilistic inference in such models, we automatically estimating the hidden quantities and thus perform our desired computation. In a sense, the rules of probability derive for us, automatically, the optimal signal processing algorithm for our desired outputs given our inputs under the model assumptions. Crucially, even though the generative model may be quite simple and may not capture all of the variability present in the data, the results of inference can still be extremely informative.

We gave several examples showing how inference in very simple generative models can be used to perform surprisingly complex speech processing tasks including denoising, source separation, pitch tracking, timescale modification and estimation of articulatory movements from audio.

6   ACKNOWLEDGMENTS

STR is funded in part by NSERC Canada, by the Premier's Research Excellence Award of Ontario and by the Canada Research Chairs Program. Thanks to Lawrence Saul and Chris Harvey for helpful discussions about the MAXVQ model. A shortened version of the first section of this chapter was presented at a special session of Eurospeech 2003 (Geneva) organized by Mazin Rahim and Rob Shapire. Work on the Segmental HMM was done in collaboration with Kannan Achan at the University of Toronto. John Hopfield first proposed the idea of a segmental model constrained to have boundaries at zero crossings.

Notes

1. An equivalent operation can be performed in the frequency domain by making a conventional spectrogram of the original signal $y(t)$ and modulating the magnitude of each short time DFT while preserving its phase: $s^{w}(\tau) = F^{-1}\{a^{w}\|F\{y^{w}(\tau)\}\|\angle F\{y^{w}(\tau)\}\}$ where $s^{w}(\tau)$ and $y^{w}(\tau)$ are the wth windows (blocks) of the recovered and original signals, α_{i}^{w} is the masking signal for subband i in window w, and $F[\bullet]$ is the DFT.

2. This can be demonstrated artificially by taking several isolated sources or noises and mixing them in a controlled way. Since the original components are known, an "optimal" set of masking signals can be computed. For example, we might set $\alpha_{i}(t)$ equal to the ratio of energy from one source in band i around times $t\pm\tau$ to the sum of energies from all sources in the same band at that time (as recommended by the Wiener filter) or to a binary version which thresholds this ratio. Constructing masks in this way is, of course, not possible when we are confronted with an unknown mixture or corrupted signal, but it can be useful for generating labeled training data for use by a statistical learning system, as discussed below.

3. Many variations on this basic theme are possible: if the final observation is obtained by stochastically selecting one of the proposed output vectors, then this becomes a "mixture of mixtures" which reduces to a large at mixture model or quantizer with a number of codebook entries equal to the product of the codebook sizes of the constituent quantizers. In Zemel's Cooperative Vector Quantization (CVQ) model (Hinton and Zemel, 1994), the proposals are combined *linearly* (either with the same coefficients across all dimensions or with different coefficients on each dimension) to produce the final output for each case. Zemel has also proposed a different model, called Multiple–Cause Vector Quantization (MCVQ) (Ross and Zemel, 2003), in MCVQ each component (dimension) of the observation vector also stochastically selects which vector quantizer is to provide its value. Each observation is then a noisy composite of the proposed values from each vector quantizer. This is like a mixture of mixtures but where each dimension makes a separate choice about which mixture to select. MAXVQ has similarities to each of MCVQ and CVQ. Unlike MCVQ, in MAXVQ the composite is not made by having each output dimension select a quantizer. Similar to CVQ there is a single fixed function which is applied to the proposed vectors from each quantizer to generate the final output. However, in CVQ this func-

tion implements a weighted sum, while in MAXVQ it implements an elementwise maximum.)

4. Much of the work described in this section was performed in collaboration with Kannan Achan at the University of Toronto.

5. A standard trick in traditional speech applications of HMMs is to use "left-to-right" transition matrices which are a special case of the type of constraints investigated in this section. However, left-to-right (Bakis) HMMs force state trajectories that are inherently one-dimensional and uni-directional whereas here we also consider higher dimensional topology and free omni-directional motion.

6. The observed symbol sequence must be "informative enough" to reveal the map structure (this can be quantified using the idea of persistent excitation from control theory).

7. Consider a network of units which compete to explain input data points. Each unit has a position in the output space as well as a position in a lower dimensional topology space. The winning unit has its position in output space updated towards the data point, but also the recent (in time) winners have their positions in topology space updated towards the topology space location of the current winner. Such a rule works well, and yields topological maps in which *nearby units code for data that typically occur close together in time*. However it cannot learn many-to-one maps in which more than one unit at different topology locations have the same (or very similar) outputs.

8. Articulator positions do not provide complete state information. For example, the excitation signal (voiced or unvoiced) is not captured by the bead locations. They do, however, provide much important information, other state information is easily accessible directly from acoustics.

9. Model structure (dimensionality and number of states) is set using cross validation.

10. An easier way (which we have used previously) to find good minima is to initialize the models using the articulatory data themselves. This does not provide as impressive "structure discovery" as annealing but still yields a system capable of inverting acoustics into articulatory movements on previously unseen test data. First, a constrained HMM is trained on just the articulatory movements, this works easily because of the natural geometric (physical) constraints. Next, we take the distribution of acoustic features (LSFs) over all times (in the training data) when Viterbi decoding places the model in a particular state and use those LSF distributions to initialize an equivalent acoustic constrained HMM. This new model is then retrained until convergence using Baum-Welch.

References

Achan, K., Roweis, S., and Frey, B., 2004. A segmental HMM for speech waveforms. *Technical Report UTML-TR-2004-001*, University of Toronto.

Blackburn, S. and Young, S., 1996. Pseudo-articulatory speech synthesis for recognition using automatic feature extraction from x-ray data. *In ICSLP 1996* **v.2**, volume 2, pages 969–972.

Brown, G.J. and Cooke, M.P., 1994. Computational auditory scene analysis. *Computer Speech and Language*, **8**.

Carreira-Perpiñán, M., 2000. Reconstruction of sequential data with probabilistic models and continuity constraints. In *Advances in Neural Information Processing Systems (NIPS)*, **volume 12**.

Cauwenberghs, G., 1999. Monaural separation of independent acoustical components. In *IEEE Symposium on Circuit and Systems (IS-CAS'99)*. IEEE.

Chennoukh, S., Sinder, D., Richard, G., and Flanagan, J., 1997. Voice mimic system using an articulatory codebook for estimation of vocal tract shape. In *Eurospeech 1997*, Rhodes, Greece.

Ephraim, Y., Malah, D., and Juang, B.H., 1989. On the application of hidden markov models for enhancing noisy speech. *IEEE Transactions on Acoustics, Speech and Signal Processing*, **37**.

Gales, M. and Young, S., 1996. Robust continuous speech recognition using parallel model combination. *IEEE Transactions on Speech and Audio Processing*, **4(5)**:352–359.

Green, P., Barker, J., Cooke, M.P., and Josifovski, L., 2001. Handling missing and unreliable information in speech recognition. In *AIS-TATS*.

Hinton, G. and Zemel, R., 1994. Autoencoders, minimum description length, and helmholtz free energy. In *Advances in Neural Information Processing Systems (NIPS)*, **volume 6**. MIT Press.

Jojic, N. and Frey, B., 2000. Topographic transformation as a discrete latent variable. In *Advances in Neural Information Processing Systems (NIPS)*, **volume 12**. MIT Press.

Logan, B. and Moreno, P., 1998. Factorial hmms for acoustic modeling. In *ICASSP*. IEEE.

Nix, D. and Hogden, J., 1999. Maximum likelihood continuity mapping: An alternative to HMMs. In *Advances in Neural Information Processing Systems (NIPS)*, **volume 11**. MIT Press.

Plante, F., Ainsworth, W.A., and Meyer, G.F., 1995. A pitch extraction reference database. In *Eurospeech*.

Ramsay, G. and Deng, L., 1994. A stochastic framework for articulatory speech recognition. *Journal of the Acoustical Society of America*, **95(5)**:2873.

Reyes, M., Raj, B., and Ellis, D., 2003. Multi-channel source separation by factorial hmms. In *ICASSP*. IEEE.

Ross, D. and Zemel, R., 2003. Multiple cause vector quantization. In *Advances in Neural Information Processing Systems (NIPS)*, **volume 15**. MIT Press.

Roweis, S., 2000. Constrained hidden markov models. In *Advances in Neural Information Processing Systems (NIPS)*, **volume 12**. MIT Press.

Roweis, S., 2001. One microphone source separation. In *Advances in Neural Information Processing Systems (NIPS)*, **volume 13**. MIT Press.

Roweis, S. and Alwan, A., 1997. Towards articulatory speech recognition. In *Eurospeech 1997*, **volume 3**, pages 1227–1230, Rhodes, Greece.

Roucos, S. and Wilgus, A.M., 1985. High quality time–scale modification for speech. In *ICASSP*. IEEE.

Schroeter, J. and Sondhi, M., 1994. Techniques for estimating vocal tract shapes from the speech signal. *IEEE Transactions on Speech and Audio Processing*, **2(1 p2)**:133–150.

Smyth, P., 1997. Clustering sequences with hidden Markov models. In G. Tesauro, D. Touretzky, and T. Leen, eds., *Advances in Neural Information Processing Systems*, **volume 9**, pages 648–654. MIT Press.

Varga, A.P. and Moore, R.K., 1990. Hidden markov model decomposition of speech and noise. In *ICASSP*, pages 845–848. IEEE.

Wan, E.A. and Nelson, A.T., 1998. Removal of noise from speech using the dual ekf algorithm. In *ICASSP*. IEEE.

Westbury, J.R., 1994. X-ray microbeam speech production database user's handbook. *Technical report*, University of Wisconsin, Madison.

Chapter 9

Signal Separation Motivated by Human Auditory Perception: Applications to Automatic Speech Recognition

Richard M. Stern
Carnegie Mellon University, Pittsburgh, PA USA
rms@cs.cmu.edu

1 INTRODUCTION

Signal separation remains one of the most challenging and compelling problems in auditory perception, and a good solution for many core signal separation problems is necessary to improve the accuracy of contemporary automatic speech recognition systems in many practical environments.

As automatic speech recognition technology is transferred from the laboratory environment into practical applications, the need to ensure robust recognition in a wide variety of acoustical environments has become increasingly apparent. While algorithms designed to cope with the effects of unknown additive noise and unknown linear filtering are plentiful in number, today's applications also demand good performance in many more difficult environments. Some of the most challenging environments for speech recognition systems today include:

- Speech in high noise, with signal-to-noise ratios (SNRs) approaching or below 0 dB
- Speech in the presence of background speech
- Speech in the presence of background music

Conventional signal processing provides only limited benefit for these problems, even today.

In this chapter I will suggest ways in which Al Bregman's huge corpus of creative research in auditory streaming and auditory scene analysis (as summarized in Bregman, 1990) can be exploited to improve the accuracy of automatic speech recognition systems. I will begin by briefly summarizing and commenting on some aspects of current state-of-the art speech recognition. I will then discuss ways in which cues that may be useful in separating speech signals for improved recognition accuracy can be extracted in ways that are based on some of the principles of auditory scene analysis.

2   ROBUST AUTOMATIC SPEECH RECOGNITION

The general topic of robust speech recognition has received a great deal of attention over the past two decades. There are many sources of acoustical distortion that can degrade the accuracy of speech recognition systems. For many speech recognition applications the two most important sources of environmental degradation are unknown additive noise (from sources such as machinery, ambient air flow, and speech babble from background talkers) and unknown linear filtering (from room acoustics, or from spectral shaping by microphones or the vocal tracts of individual speakers). Other sources of degradation include transient interference to the speech (such as doors slamming or telephones ringing), nonlinear distortion (arising from sources such as speech coding), and "co-channel" interference by individual competing talkers. Similarly, there are many approaches to robust recognition, including the use of statistical estimation of and compensation for the effects of degradation, the use of physiologically–motivated signal processing techniques that mimic processing by the human auditory system, and the use of arrays of microphones. These approaches and others are reviewed in Juang (1991); Singh *et al.* (2002a; 2002b), and Stern *et al.* (1996; 1997), among many other sources. Most research in robust recognition has been directed toward compensation for the effects of additive noise and linear filtering.

2.1   Statistical approaches to robust recognition

Figure 9.1 describes the implicit model for environmental degradation introduced by Acero and Stern (1990) and now used in many signal processing algorithms developed at CMU and elsewhere. It is assume that the "clean" speech signal $x[m]$ is first passed through a linear filter with unit sample response $h[m]$ whose output is then corrupted by uncorrelated additive noise $n[m]$ to produce the degraded speech signal $z[m]$. Under these circumstance, the goal of compensation is, in effect, to estimate the parameters characterizing the unknown additive noise and the unknown linear filter, and to apply the appropriate inverse operation. The popular approaches of spectral subtraction (as introduced by Boll, 1979) and homomorphic deconvolution (Stockham *et al.,* 1975) are special cases of this model, in which either additive noise or linear filtering effects are considered in isolation. When the compensation parameters are estimated jointly, the problem becomes a non-

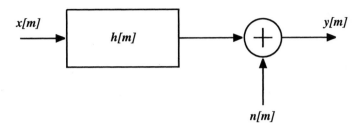

Figure 9.1. A model of environmental distortion. The effects of unknown linear filtering are modeled by the filter with unit impulse response $h[m]$ and the effects of additive unknown noise are modeled by the random process $n[m]$.

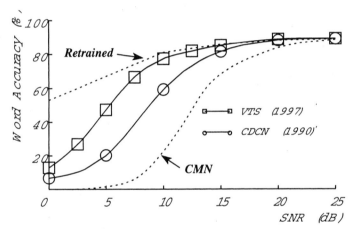

Figure 9.2. Comparison of recognition accuracies on the DARPA 5000-word Wall Street Journal task using CMN, CDCN, VTS, and complete retraining.

linear one, and can be solved using algorithms such as codeword–dependent cepstral normalization (CDCN, Acero and Stern, 1990) and vector–Taylor series compensation (VTS, Moreno *et al.*, 1996).

Figure 9.2 shows recognition accuracies obtained for a standard dictation task obtained using the CMU SPHINX-II speech recognition system for speech in broadband noise plotted as a function of signal-to-noise ratio (SNR) (Moreno *et al.*, 1996). The curve on the right represents the accuracy obtained using features derived from Mel frequency cepstral coefficients (MFCC) using cepstral mean normalization (CMN), which represents baseline performance for this particular system on this task with no particular compensation scheme used. The curve on the far left represents system performance obtained when the system is completely retrained for a particular noisy environment, which represents in a sense the upper bound in performance imposed by the particular noisy environment, given the type of signal processing and speech recognition algorithms used. The intermediate curves represent the recognition accuracy obtained using the CDCN and VTS algorithms, which were introduced in 1990 and 1997, respectively. The use of VTS provides an improvement of approximately 7 dB in SNR compared to the baseline processing. While that may not appear to be very much improvement, it can be the difference between virtually chance recognition performance and best possible performance at intermediate SNRs in the range of 5 to 10 dB, which is in important operating region.

Nevertheless, statistical parameter estimation compensation methods are not without their shortcomings. Figure 9.3 compares the *improvement* in word error rate (WER) obtained using CDCN for a similar speech recognition to that of Figure 9.2. (Results with VTS would be similar). It is expected that the improvement provided by CDCN and similar algorithms would be small at very high SNRs (because the interfering signal introduces very little degradation at those SNRs) and at very low SNRs (because the noise produces almost complete degradation no matter what form of compensation is attempted. More interesting is the performance of CDCN compensation at intermediate SNRs, where the WER is decreased by almost 50 percent with background noise, but never by more than 10 percent in the presence of background music (Raj *et al.* 1997). The failure of CDCN and similar compensation algorithms to provide meaningful compensation in the presence of background music can be attributed to several factors including the nonstationarity of background music as well as its speechlike nature. A wide variety of classical noise and channel compensation algorithms will exhibit similar deficiencies.

These difficulties with classical statistically–based approaches to robust speech recognition suggest that viable solutions to the problems of speech

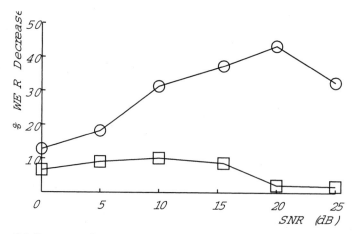

Figure 9.3. Percentage improvement provided by the CDCN algorithm for speech in the presence of white noise (circles) and background music (squares).

recognition at low SNRs and in the presence of transient and other types of time–varying interference must be based on the identification of the speech signal to be recognized, along with its explicit separation from the interfering signal or signals. This can in principle be accomplished by a number of techniques including any of several "missing–feature" approaches to noise compensation, as well as the techniques that are collectively referred to as "computational auditory scene analysis" (CASA). Researchers in CASA attempt to develop computational techniques that mimic the processes that are believed to mediate the identification and separation by humans of the separate components of a complex acoustical sound field. These approaches are discussed in the following two sections.

2.2 "Missing–feature" approaches to robust recognition

One potentially useful approach to speech recognition in the presence of the type of transient interference that is not handled well by algorithms like CDCN is the use of "missing–feature" techniques. Briefly, in missing–feature approaches, one attempts to determine which cells of a spectrogram-like time–frequency display of speech information are unreliable (or "missing") because of degradation due to noise or some other type of interference. The cells that are determined to be "missing" are either ignored in subsequent processing and statistical analysis, or they are "filled in" by optimal estimation of their putative values. While missing–feature approaches were initially motivated by similar techniques developed in image classification to deal with the problem of partially–occluded objects and have been developed by a number of research groups, it is fair to say that Martin Cooke and his colleagues at the University of Sheffield have produced the most comprehensive and widely–adopted approaches to the problem (*e.g.* Cooke *et al.*, 2001; Cooke, 2004; Brown and Palomäki, 2004).

As an example, the upper panel of Figure 9.4 shows a spectrogram of an utterance recorded in quiet. The central panel of that figure shows the same utterance after it is mixed with white noise at an SNR of 15 dB. It can be seen that the major effect of the noise is to fill in the "valleys" of the spectrogram. The lower panel of Figure 9.4 shows the same spectrogram, but the pixels that have an effective SNR of less than zero dB are indicated by dark blue solid pixels.

Figure 9.5 compares the speech recognition accuracy that is obtained using two types of missing–feature reconstruction techniques with baseline processing and simple spectral subtraction in the presence of artificially–added white Gaussian noise (upper panel) and background music derived from the DARPA Hub 4 task (lower panel), as a function of SNR (Raj *et al.*, 2004). The two missing–feature techniques that are used in these experiments,

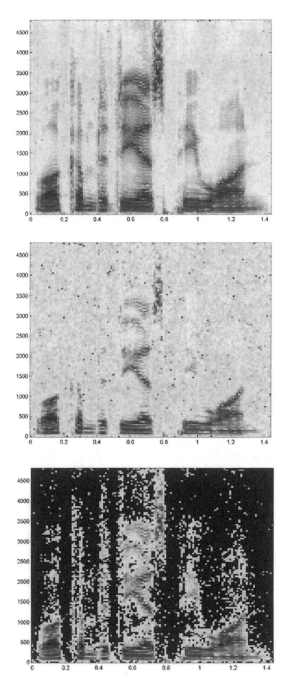

Figure 9.4. Spectrograms of speech recorded in quiet (upper panel) and subjected to artificially–added white noise with an SNR of 15 dB (central panel). Pixels that exhibit an SNR of less than zero dB are deemed missing and are depicted as solid dark regions.

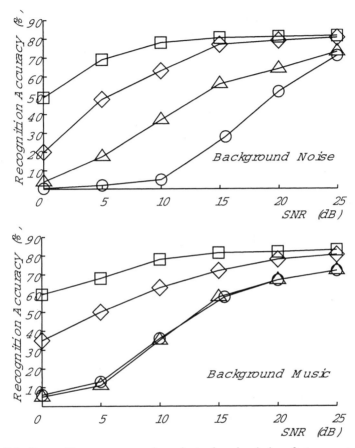

Figure 9.5. Recognition accuracy using cluster–based missing feature reconstruction (squares), covariance–based missing feature reconstruction (diamonds), simple spectral subtraction (triangles), and cepstral mean normalization only (circles) techniques for speech in the presence of white noise (upper panel) and music (lower panel) when perfect a priori information is available concerning which incoming features are "missing."

cluster–based reconstruction and covariance–based reconstruction, reconstruct the incoming feature vectors rather than modify the internal representation used by the classifier, as is more common. Recognition accuracy using missing–feature techniques can be quite good, even at low SNRs, while compensation using spectral subtraction does not improve performance at all in the presence of music. (Algorithms like CDCN and VTS would perform similarly to spectral subtraction for stimuli such as these.) Nevertheless, these results were obtained assuming perfect (or "oracle") *a priori* knowledge of which pixels in the spectrogram-like representation are "missing" and which pixels are "present" (or more accurately which pixels are damaged and which are undamaged by the effects of noise). This type of information is nor-

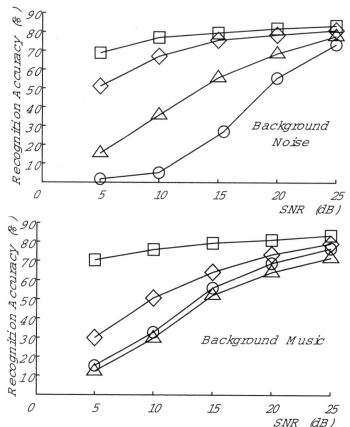

Figure 9.6. Comparison of recognition accuracy using cluster-based missing feature techniques assuming perfect "oracle" a priori knowledge of which features are missing (squares) and using blind identification of missing features (diamonds), using spectral subtraction as the basis for missing-feature decisions (triangles), and baseline processing (circles) for speech in the presence of while noise (upper panel) and music (lower panel).

mally not available in the recognition process and the blind determination of which pixels are or are not missing is in general a very difficult task.

Figure 9.6 presents a more realistic picture of the current state-of-the-art in missing–feature recognition in that it compares the recognition accuracy obtained with the missing features identified blindly using Bayesian techniques (Seltzer *et al.,* 2004) with oracle missing–feature identification, along with results obtained using spectral subtraction to obtain missing–feature decisions, and baseline processing. The most effective features used by the Bayesian classifier track the fundamental frequency of voiced speech segments and estimate the fraction of total energy in a frame that is observed at frequencies that are harmonic multiples of the fundamental. The performance obtained with blind estimation of missing features approaches that observed with perfect oracle missing–feature identification in the case of background noise. Recognition accuracy obtained using cluster–based missing–feature compensation is not quite as good in the presence of background music, but it is still quite a bit better than the accuracy obtained with statistical estimation techniques such as spectral subtraction, which do not provide any meaningful benefit at all.

2.3   Summary

While conventional techniques that compensate for the effects of additive noise and linear filtering of speech sounds can provide substantial improvement in recognition accuracy when the cause of the acoustical degradation is quasi-stationary, little improvement is observed at SNRs below approximately +5 dB. The use of techniques based on missing–feature analysis can provide substantial benefit at lower SNRs, but they are critically dependent on the ability to identify correctly which pixels actually are missing. The recognition of speech at lower SNRs, and especially speech in the presence of transient sources of interference including background speech and background music, remains a problem that is essentially unsolved at present.

3   APPLICATIONS OF AUDITORY SCENE ANALYSIS TO AUTOMATIC SPEECH RECOGNITION

Over a period of several decades, Al Bregman, his colleagues, and researchers in other groups have compiled a monumental corpus of experimental results and schematic modeling that attempt to identify ways in which the human auditory system segregates and identifies components of a complex sound field (*e.g.* Bregman, 1990; Darwin and Carlyon, 1995). While this work had originally been called "auditory streaming" by Bregman, it is now commonly known as "auditory scene analysis." The computational simulation and emulation of many of the processes identified by Bregman and his colleagues has become a popular topic of research by computer scientists and engineers in recent years, and these efforts are collectively referred to as "computational auditory scene analysis (CASA)."

Bregman *et al.* have identified many types of cues that can be used for auditory scene analysis of speech signals, including (among several others) fundamental frequency and harmonic relationships, spatial location cues, and correlated frequency and amplitude changes. In this section I will discuss some attributes about these cues and how they can be applied to improve automatic speech recognition accuracy.

3.1   Fundamental frequency and harmonic relationships

It has already been noted that pitch information can be extremely useful in the Bayesian determination of which pixels are missing in missing–feature analysis. In principle, the accurate identification and tracking of the fundamental frequency of voiced segments can be used to isolate the fundamental frequency and its harmonics from the background. In assessing the potential utility of pitch estimates as the basis for improved signal processing to achieve robust speech recognition, some key questions are how well speech can be separated from noise, how well speech signals can be separated from one another, and the extent to which this separation can improve recognition accuracy.

An informal assessment of some of these issues was made using samples of speech in the CMU Arctic database, which includes a phonetically–balanced corpus of read speech combined with electrolaryngograph (EGG) recordings collected by Komenik and Black (2003) as a resource for speech synthesis. It is quite easy to extract an accurate pitch track from the EGG recordings.

In an informal pilot study, two utterances from the Arctic database, by one male and one female speaker. The analysis and subsequent resynthesis of the speech were explored using two methods. The first approach, called synchronous heterodyne analysis (SHA), multiplies the incoming signal by a sine wave and cosine wave at the instantaneous fundamental frequency, squares the product, and sums the result over a limited time time. In the second approach, called comb–filter analysis (CFA), the speech signal is passed through a comb filter with the transfer function

$$H(z) = \frac{z^{-P}}{1 - gz^{-P}} \tag{9.1}$$

This filter has a response with sharp peaks at integer multiples of the reciprocal of the parameter P, which represents the nominal period of the signal. Varying the parameter P in accordance with the estimated fundamental frequency, and using values of 0.8 to 0.9 for the parameter g, it is possible to isolate the speech from background interference. Clearly neither SHA nor CFA provides any benefit for unvoiced segments of speech sounds or for whispered speech.

Speech in isolation was analyzed and resynthesized using the SHA and CFA methods. For both male and female speakers informal listening suggests that intelligibility is fair to good using the SHA method and good to excellent using the CFA method. Male and female speech sounds were also added together at 0 dB SNR and attempted to separate them using SHA and CFA, with the heterodyne frequency in SHA or the fundamental frequency in CFA tuned to the target speaker. Using both SHA and CFA, the separated speech of the male was fair to good in intelligibility, while the separated speech of the female was poor to fair. It is believed that this asymmetry in performance is a consequence of the differing spectral regions of male and female speech. Specifically, when the male is the target speaker, the higher fundamental frequency of the interfering female causes her speech components to be spaced farther apart in frequency, imposing a smaller amount of degradation on the upper components of the target male. Conversely, when the female is the target, the upper partials of the speech of the interfering male are relatively dense, and they are more likely to interfere with the perceptually–important lower harmonics of the speech of the target female.

While neither the SHA or CFA technique have yet been used in actual speech recognition experiments, I regard them as promising, both for speech at lower SNRs, and for speech in the presence of interfering speech and music. Again, it must be stressed that the results described above were obtained using perfect "oracle" knowledge of the fundamental frequency of the target speaker. While fundamental frequency extraction continues to be the object of a great deal of attention in recent years (*e.g.* de Cheveigné and Baskind, 2003; de Cheveigné, 2004; Kawahara *et al.,* 1999; Kawahara and Irino, 2004), pitch tracking, and especially tracking the pitch of multiple speech or music sources, remains a very difficult problem. As noted above, these techniques are not useful for unvoiced speech segments.

3.2   Spatial location cues

Sound sources arriving from different azimuths produce interaural time delays (ITDs) and interaural intensity differences (IIDs) as they arrive at the two ears. It is well known that human listeners can use spatial information to improve the intelligibility of speech in the presence of other speech or noise interference (*e.g.* Zurek, 1993). The binaural hearing mechanism can focus attention on a target speaker in a complex acoustical environment, or it can focus on the direction of arrival of the direct sound field of a target speaker in a reverberant environment. The mechanisms underlying these abilities are not completely understood, and as Zurek (1993) has noted, some improvement is to be expected simply by attending solely to the ear that is closer to the target speech source.

Most models of binaural perception (*e.g.* Colburn, 1995; Colburn and Durlach, 1978; Stern and Trahiotis, 1995, 1997) assume that peripheral auditory processing includes bandpass filtering and nonlinear rectification of the

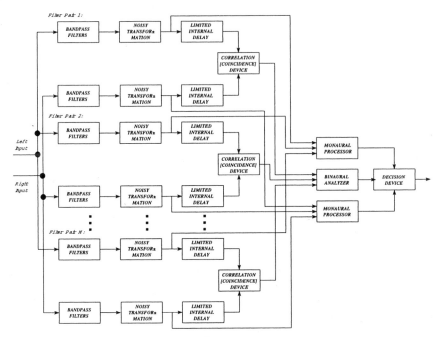

Figure 9.7. Generic model of binaural processing proposed by Colburn and Durlach.. Three of many sets of fiber pairs are depicted.

incoming sounds, followed by a cross–correlation analysis of the bandpass–filtered and rectified signals, with subsequent analysis (at least for simple stimuli) based on consideration of ITD and IID information as a function of frequency. This processing is summarized by the block diagram of Figure 9.7. Most models of binaural interaction assume that ITD information is extracted using a coincidence–analysis mechanism first proposed by Jeffress (1948). Figure 9.8 shows the putative representation of such interaural timing information in response to bandpass noise presented with an ITD of –1.5 ms with a center frequency of 500 Hz and bandwidths of 50 Hz (upper panel) and 800 Hz (lower panel). For both bandwidths the stimulus ITD can be inferred from the ridges at –1.5 ms, although with greater ambiguity in the case of the narrowband noise with only 50-Hz bandwidth (upper panel).

There is some disagreement concerning the extent to which ITD information is used by humans as the basis of signal separation. While the results of one influential study by Culling and Summerfield (1995) imply that simultaneously–presented unmodulated whispered vowels are not separated by their ITDs, more recent studies have shown that these ITD can be a useful cue in fostering identification of simultaneously–presented speech sounds when they are presented with natural amplitude and frequency modulations (Stern *et al.,* 2004). In any case, even if the auditory system does not make efficient use of

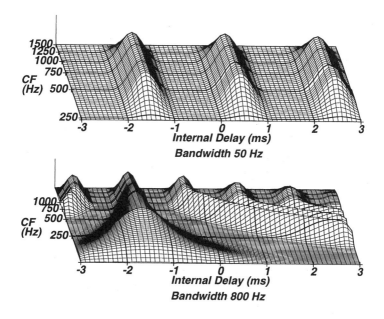

Figure 9.8. The putative response of an ensemble of Jeffress-Colburn coincidence–counting units to low–frequency bandpass noise with a center frequency of 500 Hz and an ITD of –1.5 ms. Upper panel: response to bandpass noise with a bandwidth of 50 Hz. Lower panel: response to bandpass noise with a bandwidth of 800 Hz.

interaural timing information for this task, these physical cues are still available for computational auditory processors. Many types of signal processing approaches developed in the 1980s and 1990s such as spectral subtraction did not improve human speech intelligibility when measured objectively, but they still proved to be useful for improving the accuracy of automatic speech recognition systems. Our work in this area is motivated by the belief that we should be inspired by but not limited by our knowledge of the human auditory system.

The block diagram of one system constructed some time ago that used ITD information to improve speech recognition accuracy is shown in Fig. 9.9 (Sullivan and Stern, 1993). The input signals from each of K channels are first delayed in order to compensate for differences in the acoustical path length of the desired speech signal to each microphone. (This is the same processing performed by conventional delay-and-sum beamforming.) The signals from each microphone are passed through a bank of bandpass filters with different center frequencies, passed through nonlinear rectifiers, and the outputs of the rectifiers at each frequency are correlated. (The correlator outputs correspond

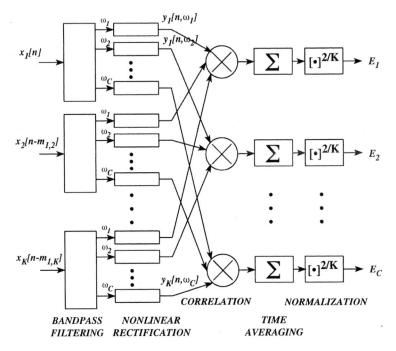

Figure 9.9. Block diagram of multi–microphone cross–correlation–based processing system.

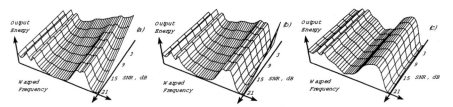

Figure 9.10. Estimates of frequency–warped spectra for the vowel segment /a/ for various SNRs using (a) 2 input channels and zero delay, (b) 2 input channels and 125-μs delay to successive channels, and (c) 8 input channels and 125-μs delay.

to outputs of the coincidence counters at the internal delays of the "ridges" in Fig. 9.8 at –1.5 ms.) The result of this operation is a form of K-dimensional cross–correlation, which reduces to a conventional cross–correlation operation for two inputs. The outputs of the multi-dimensional cross–correlation operation are considered as if they were energy estimates short–time energy estimates in each of the frequency channels, and they are subsequently converted into 12 cepstral coefficients using the cosine transform. These cepstral coefficients along with an additional coefficient representing the power of the signal are used as features for speech recognition in the conventional fashion.

Figure 9.10 demonstrates the validity of the such processing in the context of an analysis of a sample of the digitized vowel segment /a/ corrupted by artificially–added white Gaussian noise at global SNRs of 0 to +21 dB. The speech segment was presented to all microphone channels identically (to simulate a desired signal arriving on axis) and the noise was presented with linearly increasing delays to the channels (to simulate an off–axis corrupting signal impinging on a linear microphone array). The processing of such a system was simulated using 2 and 8 microphone channels, and time delays for the masking noise of 0 and 125 μs to successive channels.

The curves of Fig. 9.10 describe the effect of SNR, the number of processing channels, and the delay of the noise on the spectral profiles of a sample of the vowel segment /a/. (The frequency representation for the vowel is warped in frequency according to the nonlinear spacing of the auditory filters.) The upper panel summarizes the results that are obtained using 2 channels with the noise presented with zero delay from channel to channel (which would be the case if the speech and noise signals arrive from the same direction). Note that the shape of the vowel, which is clearly defined at high SNRs, becomes almost indistinct at the lower SNRs. The center and lower panels show the results of processing with 2 and 8 microphones, respectively, when the noise is presented with a delay of 125 μs from channel to channel (which corresponds to a off–axis source location for typical microphone spacing). As the number of channels increases from 2 to 8, the shape of the vowel segment in Figure 9.10 becomes much more invariant to the amount of noise present. In general, it was found in these experiments that the benefit to be expected from increases sharply as the number of microphone channels is increased. It was also observed (unsurprisingly) that the degree of improvement increases as the simulated directional disparity between the desired speech signal and the masker increases. It was concluded from these pilot experiments that the cross–correlation method described can provide very good robustness to off–axis additive noise, and in practice, this approach did provide a moderate benefit over the recognition accuracy obtained using conventional delay-and-sum processing (Sullivan and Stern, 1993).

These studies, which were conducted in the early 1990s, were not continued because of the lack of available computational resources at that time. It is believed that further improvements are likely to be obtained once greater attention is paid to the nature of the bandpass filtering, within–channel nonlinearities, and correlation operations. Correlation–based approaches such as these can be applied to unvoiced as well as to voiced segments of speech, and other research groups as well as ours are experiencing success in the application of information about interaural differences to improved speech recognition accuracy (*e.g.* Palomäki *et al.,* 2004, Wang, 2004).

A final item of note with regard to spatial processing is that high levels of reverberation are extremely detrimental to recognition accuracy. Frame-based compensation strategies such as those discussed in Sec. 2.1 fail because the effects of reverberation are generally spread over multiple analysis frames. In addition, traditional adaptive filtering methods, which have also been considered for this purpose, depend on the statistical independence of target and masker. Since in reverberant environments, the "noise" consists of reflected and attenuated copies of the target speech signals, noise–canceling adaptive filter strategies (such as those that use the LMS algorithm) are generally not effective. Sub-band processing in a similar fashion has been somewhat effective in reverberation, but it has not yet been applied to a wide range of problems. It is expected that techniques based on auditory perception and physiology, missing–feature recognition, and CASA techniques should be more effective in characterizing and ultimately ameliorating the effects of reverberation.

3.3   Correlated frequency and amplitude changes

Although cues based on fundamental frequency and sound source location are potentially extremely valuable in separating signals for automatic speech recognition, pitch cues are ineffective for unvoiced segments and location multichannel recordings with location information are not always available. Even with just a single channel, unvoiced speech segments, and/or imperfect pitch estimates, we expect to be able to separate multiple sound sources using by extracting and clustering sounds according to small ("micro") modulations in frequency and amplitude. The use of such physical cues for sound separa-tion and auditory scene analysis has been supported by many psychoacoustical studies in recent years (*e.g.* Bregman, 1990; Darwin and Carlyon, 1995).

In an influential demonstration, John Chowning showed in the early 1980s that the perceptual salience of correlated frequency modulation is a highly salient factor in segregating and fusing components of a complex signal or signals. As described in detail in Bregman's (1990) treatise, an initial com-plex tone is presented that combines three sine waves at (for example) 300, 400, and 500 Hz. Taken as individual sine waves, these three frequencies form a major triad in the second inversion, but the components are more likely to be heard as the third through fifth harmonics of a fused complex tone with fundamental frequency 100 Hz. Next, the three sine waves are replaced by three sets of 10 harmonics at integer multiples of 300, 400, and 500 Hz, with spectral envelopes that are derived from three different vowel sounds. Again this signal is perceived as a complex tone with fundamental frequency 100 Hz, but with a sharper timbre because of the presence of 27 additional upper–

frequency partials. Finally, separate frequency modulations at approximately 4.5, 5, and 5.5 Hz, are added to the 10-harmonic complex tones with fundamental frquencies 300, 400, and 500 Hz, respectively. Once the frequency modulation is applied, the harmonics associated with each fundamental frequency segregate from one another and become easily perceived as three separate complex tones with fundamental frequencies of 300, 400, and 500 Hz. In informal listening we have noted that the segregation according to frequency modulation is dramatic and easy to perceive using the signal parameters as described in Bregman (1990). Signals that are constructed by amplitude modulating the partials in a similar fashion do not produce nearly as dramatic a perceptual segregation according to modulation frequency as is observed for frequency modulation.

My research group is currently developing ways to develop through computational means a form of signal separation based on the identification of common locations in a time–frequency display like a spectrogram that exhibit covarying amplitude and frequency modulation. This is a difficult task because of the need to achieve sharp frequency resolution while allowing for temporal fluctuations, but the ability of the human auditory system to accomplish these tasks remains a powerful existence proof and motivation to move this work forward.

4   SUMMARY

It is believed that computational auditory approaches are potentially extremely useful in ameliorating some of the most difficult speech recognition problems, specifically the recognition of speech presented at low SNRs, speech masked by other speech, speech masked by music, and speech in highly reverberant environments. The solution to these problems using CASA techniques is likely to depend on the ability to develop several key elements of signal processing, including the reliable detection of fundamental frequency for isolated speech and for multiple simultaneously–presented speech sounds, the reliable detection of modulations of amplitude and frequency in very narrowband channels, and the development of across–frequency correlation approaches that can identify frequency bands with coherent micro–activity as they evolve over time. I am extremely optimistic that effective solutions for these problems are within reach in the near future.

ACKNOWLEDGMENTS

The author is extremely grateful to Dan Ellis, DeLiang Wang, and especially Pierre Divenyi for organizing the NSF Workshop on Speech Separation and to the National Science Foundation's Human Language and Communication Program and the Mitsubishi Electric Research Labs for sponsoring it. The work described was sponsored by the Space and Naval Warfare Systems Center, San Diego, under Grant No. N66001-99-1-8905. The content of the information in this publication does not necessarily reflect the position or the policy of the US Government, and no official endorsement should be inferred.

References

Acero, A., and Stern, R. M., 1990, "Environmental robustness in automatic speech recognition," *Proc. ICASSP*, Albuquerque, New Mexico.

Boll, S. F., 1979, "Suppression of acoustic noise in speech using spectral subtraction," *IEEE Trans. Acoustics, Speech and Signal Processing,* **27:** 113–120.

Bregman, A. S., 1990, *Auditory Scene Analysis: The Perceptual Organization of Sound,* Cambridge: MIT Press, Cambridge.

Brown, G. J., and Palomäki, K., 2004, "Techniques for speech processing in noisy and reverberant conditions," this volume.

Colburn, H. S., 1995, "Computational models of binaural processing," in *Springer Handbook of Auditory Research: Auditory Computation*, H. L. Hawkins, T. A. McMullen, A. N. Popper, and R. R. Fay, eds. New York: Academic Press, pp. 332–400.

Colburn, H. S., and Durlach, N. I., 1978, "Models of binaural interaction," in *Handbook of Perception*, E. C. Carterette and M. P. Friedman, eds., Academic Press, New York, pp. 467–518.

Cooke, M., Green, P. Josifovski, L., and Vizinho, A, 2001, "Robust automatic speech recognition with missing features and unreliable acoustic data," *Speech Communication,* **34:** 267–285.

Cooke, M., 2004, "Making sense of everyday speech: A glimpsing account," this volume.

Culling, J. F., and Summerfield, Q., 1995, "Perceptual separation of concurrent speech sounds: Absence of across–frequency grouping by common interaural delay," *J. Acoust. Soc. Am.* **98:** 785–797.

Darwin, C. J., and Carlyon, R. P., 1995, "Auditory grouping," in *Handbook of Perception and Cognition, Vol. 6: Hearing*, B. C. J. Moore., ed. New York: Academic Press, pp. 347–386.

de Cheveigné, A., 2004, "The cancellation principle in acoustic scene analysis," this volume.

de Cheveigné, A., and Baskind, A., 2003, "F0 estimation of one or several voices," *Proc. Eurospeech,* pp. 833–836.

Jeffress, L. A., 1948, "A place theory of sound localization," *J. Comparative and Physiological Psychology* **41:** 35–39.

Juang, B.-H., 1991, "Speech recognition in adverse environments," *Computer Speech and Language*, **5**: 275–294.

Lindemann, W., 1986, "Extension of a binaural cross–correlation model by contralateral inhibition. I. Simulation of lateralization for stationary signals," *J. Acoust. Soc. Am.* **80**, 1608–1622.

Kawahara, H., and Irino, T., 2004, "Underlying principles of a high–quality speech manipulation system STRAIGHT and its application to speech segregation," this volume.

Kawahara, H., Matsuda-Katsuse, I., and de Cheveigné, A., 1999, "Restructuring speech representations using a pitch adaptive time–frequency smoothing and an instantaneous–frequency–based F0 extraction: Possible role of a repetitive structure in sounds," *Speech Communication, 27*: 175–185.

Komenek, J., and Black, A., 2003, *CMU_ARCTIC Databases,*
http://www.festvox.org/cmu_arctic.

Palomäki, K. J., Brown, G. J., and Wang, D. L., 2004, "A binaural processor for missing data speech recognition in the presence of noise and small–room reverberation," *Speech Communication* (accepted for publication).

Moreno, P. J., Raj, B., and Stern, R. M., 1996, "A vector taylor series approach for environment–independent speech recognition," *Proc. ICASSP*, Atlanta, Georgia.

Raj, B., Parikh, V. N., and Stern, R. M., 1997, "The effects of background music on speech recognition accuracy," *Proc. ICASSP*, Munich, Germany.

Raj, B., Seltzer, M. L., and Stern, R. M., 2004, "Reconstruction of missing features for robust speech recognition," *Speech Communication Journal* (accepted for publication).

Seltzer, M. L., Raj, B., and Stern, R. M., 2004, "A Bayesian Framework for Spectrographic Mask Estimation for Missing Feature Speech Recognition," *Speech Communication Journal* (accepted for publication).

Singh, R., Stern, R. M. and Raj, B., 2002a, "Signal and feature compensation methods for robust speech recognition," Chapter in *CRC Handbook on Noise Reduction in Speech Applications*, Gillian Davis, ed., CRC Press, Boca Raton.

Singh, R, Raj, B. and Stern, R. M., 2002b, "Model compensation and matched condition methods for robust speech recognition," Chapter in *CRC Handbook on Noise Reduction in Speech Applications*, Gillian Davis, ed. CRC Press, Boca Raton.

Stern, R. M., Acero, A. Liu, F.-H. Liu, and Oshima, Y., 1996, "Signal processing for robust speech recognition," Chapter in *Automatic Speech and Speaker Recognition*, C.-H. Lee, F. Soong, and K. Paliwal, eds., Kluwer Academic Publishers, Boston, pp. 351–378.

Stern, R. M., Raj, B. and Moreno, P. J., 1997, "Compensation for environmental degradation in automatic speech recognition," *Proc. ETRW on Robust Speech Recognition for Unknown Communication Channels*, Pont-au-Mousson, France, pp. 33–42.

Stern, R. M., and Trahiotis, C., 1995, "Models of binaural interaction," in *Handbook of Perception and Cognition, Volume 6: Hearing*, B. C. J. Moore., ed., Academic Press, New York, pp. 347–386.

Stern, R. M., and Trahiiotis, C., 1996), Models of Binaural Perception," in *Binaural and Spatial Hearing in Real and Virtual Environments*, R. Gilkey and T. R. Anderson, Eds. New York: Lawrence Erlbaum Associates, pp. 499–531.

Stern, R.. M., Trahiotis, C., and Ripepi, A. M., 2004, "Some conditions under which interaural delays foster identification," in *Dynamics of Speech Production and Perception*, G. Meyer and P. Divenyi,eds., IOP Press, Amsterdam: IOP Press (in press).

Stockham, T. G., Cannon, T. M., and Ingebretsen, R. B., 2004, "Blind Deconvolution Through Digital Signal Processing," *Proc. IEEE*, **63**: 678–692/

Sullivan, T. M., and Stern, R. M., 1993, "Multi–Microphone Correlation–Based Processing for Robust Speech Recognition," *Proc. ICASSP*, Minneapolis, Minnesota.

Wang, D., 2004, "On the use of ideal binary time–frequency masks for CASA," this volume.

Zurek, P. M., "Binaural Advantages and Directional Effects in Speech Intelligibility, in *Acoustical Factors Affecting Hearing Performance II,* G. A. Studebaker and I. Hochberg, Eds. Boston: Allyn and Bacon, 1993.

Chapter 10

Speech Segregation Using an Event–synchronous Auditory Image and STRAIGHT

Toshio Irino
Faculty of Systems Engineering, Wakayama University, Wakayama, JAPAN
irino@sys.wakayama-u.ac.jp

Roy D. Patterson
CNBH, Physiology Department, Cambridge University, Cambridge, UK
roy.patterson@mrc-cbu.cam.ac.uk

Hideki Kawakhara
Faculty of Systems Engineering, Wakayama University, Wakayama, JAPAN
kawahara@sys.wakayama-u.ac.jp

1   INTRODUCTION

Speech segregation is an important aspect of speech signal processing, and many segregation systems have been proposed (Parsons, 1976, Lim *et al.*, 1978). Most of the systems are based on the short–time Fourier transform, or a sinusoid model, and they segregate sounds by extracting harmonic components located at integer multiples of the fundamental frequency. It is, however, difficult to extract truly harmonic components when the estimation of the fundamental frequency is disturbed by concurrent noise, the error increases in proportion to the harmonic number.

We have developed an alternative method for speech segregation based on the Auditory Image Model (AIM) and a scheme of event–synchronous processing. AIM was developed to provide a reasonable representation of the "auditory image" we perceive in response to sounds (Patterson *et al.*, 1992, 1995). We have also developed an "auditory vocoder" (Irino *et al.*, 2003a,b, 2004) for resynthesizing speech from the auditory image using an existing, high–quality vocoder, STRAIGHT (Kawahara *et al.*, 1999). The auditory representation preserves fine temporal information, unlike conventional window–based processing, and this makes it possible to do synchronous speech segregation. We also developed a method to convert F0 into event times and demonstrated the potential of the method in low SNR conditions.

This paper presents a procedure for segregating speakers using an event–synchronous auditory image with two different methods of resynthesizing the speech of the individual speakers. We also show that the system is robust in the presence of errors in fundamental frequency estimation which are unavoidable.

2   AUDITORY VOCODER

We propose two versions of the auditory vocoder as shown in Figs 10.1 and 10.2. They involve the Auditory Image Model (AIM) (Patterson *et al.*, 1995), a robust F0 estimator (e.g., Nakatani and Irino, 2002), and a synthesis module based either on STRAIGHT (Fig. 10.1) or an auditory synthesis filterbank (Fig. 10.2). The speech segregation is performed in the event–synchronous version of AIM which is common in the two versions. In section 2.1, we describe the event–synchronous procedure to enhance speech representations and a method for converting F0 to event times. In section 2.2, we describe two procedures for resynthesizing signals from the auditory image.

2.1   Event based Auditory Image Model

AIM performs its spectral analysis with a gammatone or gammachirp filterbank on a quasi-logarithmic (ERB) frequency scale. The output is half–wave rectified and compressed in amplitude to produce a simple form of Neural Activity Pattern (NAP). The NAP is converted into a Stabilized Auditory Image (SAI) using a strobe mechanism controlled by an event detector. Basically, it calculates the times between neural pulses in the auditory nerve and constructs an array of time–interval histograms, one for each channel of the filterbank.

2.2   Event detection

An event–detection mechanism was introduced to locate glottal pulses accurately for use as strobe signals. The upper panel of Fig. 10.3 shows about 30 ms of the NAP of a male vowel. The abscissa is time in ms, the ordinate is the center frequency of the gammatone auditory filter in kHz. The mechanism identifies the time interval associated with the repetition of the neural pattern and the interval is the fundamental period of the speech at that point. Since the NAP is produced by convolution of the speech signal with the auditory filter and the group delay of the auditory filter varies with center frequency, it is necessary to compensate for group delay across channels when estimating event timing in the speech sound. The middle panel shows the NAP after

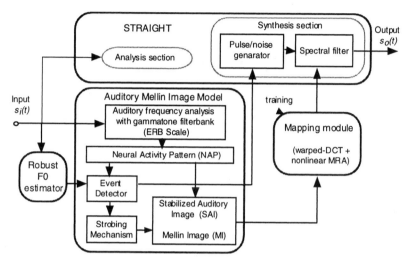

Figure 10.1. Configuration of the auditory vocoder: The gray areas show the processing path used for parameter estimation. Once the parameters are fixed, the path with the black arrows is used for resynthesizing speech signals.

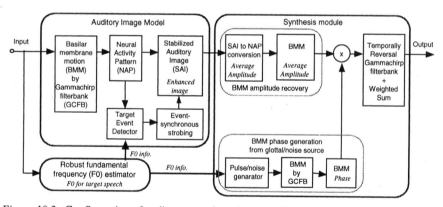

Figure 10.2. Configuration of auditory vocoder using an auditory synthesis filterbank. There is no training path.

group delay compensation, the operation aligns the responses of the filters in time to each glottal pulse. The solid line in the bottom panel shows the temporal profile derived by summing across channels in the compensated NAP in the region below 1.5 kHz. The peaks corresponding to glottal pulses are readily apparent in this representation.

We extracted peaks locally using an algorithm similar to adaptive thresholding (Patterson *et al.*, 1995). The dotted line is the threshold used to identify peaks, indicated by circles and vertical lines. After a peak is found, the thresh-

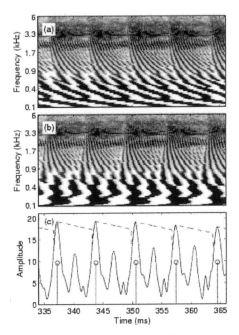

Figure 10.3. Auditory event detection for clean speech. From Irino, *et al.* (2003a), © IEEE
2003

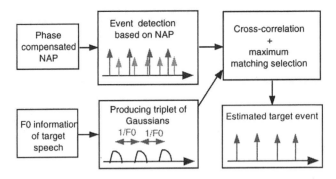

Figure 10.4. Selection of Event time by F0 information.

old decreases gradually to locate the next peak. We also introduced a form of
prediction to make the peak detection robust. The threshold is reduced by a
certain ratio when the detector does not find activity at the expected period,
defined as the median of recent periods. This is indicated by the sudden drop
in threshold towards the end of each cycle.

Tests with synthetic sounds confirmed that this algorithm works sufficiently well when the input is clean speech. It is, however, difficult to apply this method under noisy conditions particularly when the SNR is low. So, we enlisted the F0 information to improve event time estimation.

2.2.1   A robust F0 estimator for event detection

It is easier to estimate fundamental frequency, F0, than event times for a given SNR. The latest methods for F0 estimation are robust in low SNR conditions and can estimate F0 within 5% of the true value in 80% of voiced speech segments in babble noise, at SNRs as low as 5 dB (Nakatani and Irino, 2002). So, we developed a method of enhancing event detection with F0 information.

Figure 10.4 shows a block diagram of the procedure. First, candidate event times are calculated using the event detection mechanism described in the previous section. The half–life of the adaptive threshold is reduced to avoid missing events in target speech. The procedure extracts events for both the target and background speech. Then we produce a temporal sequence of the pulses located at the candidate times.

For every frame step (e.g., 3 ms), the value of F0 of the target speech in that frame is converted into temporal function consisting of a triplet of gaussian functions with a periodicity of 1/F0. The triplet function is, then, cross–correlated with the event pulse sequence to find the best matching lag time. At this best point, the temporal sequence and the triplet of gaussians are multiplied together so as to derive a value similar to a likelihood for each event. The likelihood–like values are accumulated and applied to the thresholding with an arbitrary value to derive estimates of the event times for the target speech signal.

2.2.2   Event synchronous strobed temporal integration

Using the target event times, we can enhance the auditory image for the target speech. Figure 10.5a shows a schematic plot of a NAP after group delay compensation for a segment of speech with concurrent vowels. The target speech with a 10-ms glottal period is converted into the black activity pattern, while the background speech with a 7-ms period is converted into the gray pattern. For every target event time (every 10 ms), the NAP is converted into a two-dimensional auditory image (AI) as shown in Fig. 10.5b. The horizontal axis of the AI is time–interval (TI) from the event time, the vertical axis is the same as the NAP. As shown in Fig. 10.4b, the activity pattern for the target speech always occurs at the same time–interval, whereas that for the background speech changes position. So, we get a "stablized" version of the target speech pattern in this "instantaneous" auditory image (AI).

Temporal integration is performed by weighted averaging of successive frames of the instantaneous AI as shown in Fig. 10.5c, and this enhances the

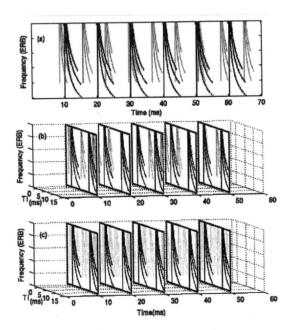

Figure 10.5. Event synchronous strobes for concurrent speech.

pattern for the target speech relative to that for the background speech. In this way, the fine structure of the target pattern is preserved and stabilized when the event detector captures the glottal event of the target speech correctly. The weighted averaging is essentially equivalent to conventional STI where the weighting function is a ramped exponential with a fixed half–life. In the following experiment, we used a hanging window spanning five successive SAI frames, although tests indicated that the window shape does not have a large effect as long as the window length is correct.

2.3   Synthesis procedure

We can now produce an auditory image in which the pattern of the target speech is enhanced. It only remains to resynthesize the target speech from their auditory representations.

2.3.3   Resynthesis using STRAIGHT

Figure 10.1 includes a synthesis method using STRAIGHT and a mapping module between the auditory image and a time–frequency representation based on the STFT. The method is described by Irino *et al.* (2003a,b). Briefly, each frame of the 2-D stabilized auditory image (SAI) is converted into a Mellin Image (MI), and then the MI is converted into a spectral distribution using a warped-DCT. The mapping function between the MI and the warped-DCT is constructed using a non-linear multiple regression analysis (MRA). Fortunately, it is possible to determine the parameter values for the non-linear MRA in advance, using clean speech sounds and the analysis section of STRAIGHT as indicated in Fig. 10.1 gray arrows. The speech sounds are, then, reproduced using a spectral filter excited with the pulse/noise generator shown in Fig. 10.1 in the synthesis section of STRAIGHT. The pulse/noise generator is controlled by the robust F0 estimator in the left box.

This version works fairly well and is expected to capture fine structure of speech signal in the SAI and MI. But it is rather complex and inflexible because training is required to establish the mapping function. Moreover, the quality of the resynthesized sounds was not always good despite the use of STRAIGHT. So, we rebuilt the synthesis procedure using a channel vocoder concept and introduced an approximation in the mapping function.

2.3.4 Auditory channel vocoder

Figure 10.2 shows the revised version based on an auditory filterbank and channel vocoder (Gold and Radar, 1967). The channel vocoder consisted of a bank of band–limited filters and pulse/noise generators controlled by previously extracted information about fundamental frequency and voicing. There were two banks of gammachirp/gammatone filters for analysis and synthesis because of the temporal asymmetry in the filter response. But the synthesis filterbank is not essential when we compensate for the phase lag of the analysis filter. The amplitude information of the Basilar Membrane Motion (BMM) is recovered from the SAI via the NAP as described in the next section. The phase information of the BMM is derived by the filterbank analysis of the signal from a pulse/noise generator. We employed the generator in STRAIGHT since it is known to produce pulses and noise with generally flat spectra, and it does not affect the amplitude information. The amplitude and phase information is combined to produce the real BMM for resynthesis.

2.3.5 Mapping function from SAI to NAP

Mapping from the SAI to the NAP required one more enhancement. In the previous version, we took the fine temporal structure in both the SAI and NAP into account when designing the mapping function. As a result, it required a non-linear mapping between multiple inputs and a single output. In the current study, we assumed that the frequency profile of the SAI provides a reasonable representation of the main features of the target speech when we restrict the range of integration in the appropriate way. As shown in Fig. 10.5c, the target feature is concentrated in the time–interval dimension when

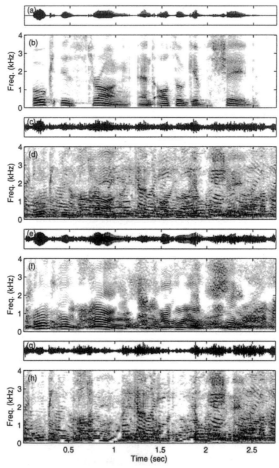

Figure 10.6. Waveform and spectrum when the F0 estimation error is 5%.(a) (b) original sound, (c)(d) mix of sound and babble noise, (e)(f) signals segregated by the proposed method, and (g)(h) by a comb–filter method.

the filter center frequency, f_c, is high, and it overlaps the next cycle when the center frequency is low. The activity of the interfering speech appears at a low level in the rest of the SAI. So, we limited the range of integration $[T_{min}, T_{max}]$ on the time–interval axis to

$T_{min} = -1$ and $T_{max} = \min(5, \max(1.5, 2f_c))$ (ms).

T_{min} is constant. T_{max} is 5 ms when f_c is low and 1.5 ms when f_c is high. We applied a weighting function to produce a flat NAP from a regular SAI. The resulting synthetic sounds were reasonably good. Moreover, the assumption simplifies the procedure. It would probably be possible to enhance the sound quality using the fine structure in the SAI.

3   EXPERIMENTS

Many segregation procedures require precise F0 information for effective segregation. This is unrealistic because F0 estimation is never error–free and natural vocal vibration contains jitter. So it is important to evaluate how robust the system is to jitter and errors in F0 estimation. The new event finding procedure produces performance that is the same, or a little better, than the previous version. It is easier to listen to the target sound than the mix of target speech and distractor speech, or the mix of target speech and babble noise.

We compared the new method to a comb–filter method using sound mixtures where F0 is 5% more than the original value. Figures 10.6a and 10.6b show the original, isolated target speech and its spectrogram. The target sound was mixed with babble noise at an SNR of 0 dB to produce the wave in Fig. 10.6c. At this level, a listener needs to concentrate to hear the target speech in the distracter. The spectrogram in Fig. 10.6d shows that the features of the target speech are no longer distinctive. The target speech (Fig. 10.6e) extracted by the auditory vocoder was distorted but entirely intelligible, whereas the distractor speech was decomposed into a non–speech sound which greatly reduced the interference. The spectrogram in Fig. 10.6f shows that the vocoder resynthesis restores the harmonic structure of the target speech but not the distractor speech, e.g., compare the restoration of harmonic structure just after 800 ms with the lack of harmonic restoration in the region just before 800 ms.

For comparison, we applied a comb–filter method based on the STFT to the same sound. The target speech extracted in this way sounds low–pass filtered and the babble noise is relatively strong. The target speech wave (Fig. 10.6g) is noisier than the wave in Fig. 10.6e. The spectrogram in Fig. 10.6f shows that much of the harmonic structure is lost by the comb–filter method, particularly the higher harmonics. This follows directly from the fact that the error increases in proportion to harmonic number. It is a fundamental defect.

4   CONCLUSIONS

We have presented methods to segregate concurrent speech sounds using an auditory model and a vocoder. Specifically, the method involves the Auditory Image Model (AIM), a robust F0 estimator, and a synthesis module based either on STRAIGHT or an auditory synthesis filterbank. The event–synchronous procedure enhances the intelligibility of the target speaker in the presence of concurrent background speech. The resulting segregation performance is better than with conventional comb–filter methods whenever there are errors in fundamental frequency estimation as there always are in real concurrent speech. Test results suggest that this auditory segregation method has potential for speech enhancement in applications such as hearing aids.

5   ACKNOWLEDGMENTS

This work was partially supported by a project grant from the faculty of systems engineering of Wakayama University, by the Ministry of Education, Science, Sports and Culture, Grant-in-Aid for Scientific Research (B), 15300061, 2003, and by the UK Medical Research Council (G9901257).

References

Gold, B. and Rader, C. M., 1967, The Channel Vocoder, *IEEE Trans. Audio and Electroacoustics*, **Vol. AU-15**, 148–161.

Irino, T., Patterson, R.D., and Kawahara, H., 2004, Speech segregation using an auditory vocoder with event–sychronous enhancements, Proc *18th International Congress on Acoustics (ICA 2004)*, vol. **4**, pp. 3025–3028, Kyoto, Japan.

Irino, T., Patterson, R.D., and Kawahara, H., 2003a, Speech segregation using event synchronous auditory vocoder, in *Proc. IEEE ICASSP 2003*, Hong Kong.

Irino, T., Patterson, R.D., and Kawahara, H., 2003b, Speech segregation based on fundamental event information using an auditory vocoder, *Proc. 8th Euro Conf. on Speech Comm. and Tech.* (Eurospeech 2003, (Interspeech 2003)), 553–556, Geneva, Switzerland.

Kawahara, H., Masuda-Katsuse, I., and de Cheveigne, A., 1999, Restructuring speech representations using a pitch–adaptive time–frequency smoothing and an instantaneous–frequency–based F0 extraction: Possible role of a repetitive structure in sounds, *Speech Comm*, **27**, 187–207.

Lim, J.S., Oppenheim, A.V., and Braida, L.D., 1978, Evaluation of an adaptive comb filtering method for enhancing speech degraded by white noise addition, *IEEE, Trans. ASSP*, **ASSP-26**, 354–358.

Nakatani, T. and Irino, T., 2002, Robust fundamental frequency estimation against background noise and spectral distortion, *ICSLP 2002*, 1733–1736, Denver, Colorado.

Parsons, T.W., 1976, Separation of speech from interfering speech by means of harmonic selection, *J.Acoust. Soc.Am.*, **60**, 911–918.

Patterson, R.D., Allerhand, M., and Giguere, C., 1995, Time–domain modelling of peripheral auditory processing: a modular architecture and a software platform, *J. Acoust. Soc. Am.*, **98**, 1890–1894. http://www.mrc-cbu.cam.ac.uk/cnbh/

Patterson, R.D., Robinson, K., Holdsworth, J., McKeown, D., Zhang, C., and Allerhand, M., 1992, Complex sounds and auditory images, In: *Auditory physiology and perception, Proc. of the 9h Internat. Symposium on Hearing*, Y. Cazals, L. Demany, K. Horner (eds), Pergamon, Oxford, 429–446.

Chapter 11

Underlying Principles of a High–quality Speech Manipulation System STRAIGHT and Its Application to Speech Segregation

Hideki Kawahara
Faculty of Systems Engineering Wakayama University, Wakayama, JAPAN
kawahara@sys.wakayama-u.ac.jp

Toshio Irino
Faculty of Systems Engineering Wakayama University, Wakayama, JAPAN
irino@sys.wakayama-u.ac.jp

1   INTRODUCTION

Human performance in auditory perceptual tasks has been investigated by many researchers. Psychophysical methods have been successfully applied to measure various aspects using elementary sound stimuli, and there are still continuing efforts to form a growing body of scientific knowledge in this field. Speech perception also has been investigated extensively, especially since the advent of electrical engineering. However, additional methodologies need to be developed to test all aspects of human performance, especially performance in natural everyday situations.

A human operate with highly nonlinear systems, thus it is generally difficult to understand its performance and functions only from responses to idealized (simplified or synthetic) stimuli. A better way of dealing with such complex nonlinear systems is to use ecologically relevant stimuli. In spite of this conceptual advantage, there has been an obvious difficulty for this approach to be feasible: the controllability of test stimuli. Usually, ecologically relevant stimuli, in other words, natural stimuli, do not allow precise control of physical parameters. This is a problem, because it is a prerequisite for scientific research. In response, a very high–quality speech manipulation system STRAIGHT (Kawahara *et al.,* 1999a) was developed to provide a means of resolving this apparent contradiction.

2   UNDERLYING PRINCIPLES OF STRAIGHT

STRAIGHT is basically a channel vocoder based on F0 adaptive procedures. It is worthwhile to briefly introduce STRAIGHT (Speech
Transformation and Representation based on Adaptive Interpolation of
weiGHTed spectrogram), which was originally designed for speech perception research using a source filter architecture. The key concept of
STRAIGHT is to represent signals in terms of parameters that embody essential aspects of auditory perception. In other words, special representations are
sought for tone, noise and transient. For tone, a time–frequency representation
that does not suffer from interference caused by periodic excitation is introduced. In the first stage of analysis, F0 adaptive design of a complementary
set of time windows effectively eliminates temporal variations in the power
spectrum estimate. Then, a spline–based F0 adaptive smoothing in the frequency domain eliminates variations due to harmonic structure. For transient,
fixed–point–based methods to extract and to represent auditory events are
introduced.

The procedures used in STRAIGHT are grouped into three subsystems: a
source information extractor, a smoothed time–frequency representation
extractor, and a synthesis engine consisting of an excitation source and a time
varying filter. Underlying principles and implementation of the second and
the third components are given in the following sections.

2.1   Separating source and filter information

Separating speech information into mutually independent filter parameters
and source parameters is important for enabling flexible speech manipula-
tions. The key idea for selectively eliminating interferences caused by
periodicity is based on a new interpretation of the role of periodic excitation
of speech. Periodic excitation can be interpreted as a systematic sampling
operation of the underlying time–frequency representation of speech that
reflects smooth movement of the articulatory organs. This sampled information can be used to reconstruct the underlying time–frequency representation
by interpolating sampled points. This two-dimensional interpolation is then
reduced to a one-dimensional smoothing operation.

The need for time–domain smoothing is eliminated by using an F0 adaptive complimentary time–window pair. The primary window is an F0 adaptive
window made from an isometric Gaussian window and an F0 adaptive Bar-
tlett window,

$$w_p(t) = \exp\left(-\pi \frac{t^2}{t_0^2}\right) * h\left(\frac{t}{t_0}\right)$$

$$h(t) = \begin{cases} 1 - |t| & |t| \leq 1 \\ 0 & \text{otherwise,} \end{cases} \quad (1)$$

where * represents convolution. The corresponding complimentary window is made from this window by introducing the following modulation.

$$w_c(t) = w_p(t) \sin\left(\pi \frac{t}{t_0}\right). \quad (2)$$

The reduced interference spectrogram is calculated with the following equation with a numerically optimized mixing coefficient ξ. A power spectrum with reduced phase interference, $P_r(\omega,t)$, is represented as a weighted squared sum of the power spectra, $P_c(\omega,t)$ and $P_o(\omega,t)$, using this compensatory window and the original time window, respectively.

$$P_r(\omega,t) = \sqrt{P_o(\omega,t) + \xi P_c(\omega,t)}. \quad (3)$$

The procedure described above eliminates needs of time–domain interpolation; this interpolation is only necessary in the frequency domain. F0 adaptive spectral smoothing based on a cardinal B-spline basis function removes interferences due to signal periodicity in the spectral slice. It should be noted that such smoothing operation needs to compensate for the over-smoothing effects due to time–windowing. Details are given in the literature (Kawahara *et al.,* 1999a). Figures 11.1, 11.2 show how interferences in a standard F0 adaptive spectrogram are systematically removed.

In the synthesis part of STRAIGHT, the filtering component is implemented as the minimum phase impulse response calculated from the smoothed time–frequency representation through several stages of FFTs. This FFT–based implementation enables source F0 control with a finer frequency resolution than that determined by the sampling interval of the speech signal. This implementation also enables suppression of "buzz–like" timbre, which is commonly found in conventional pulse excitation, by introducing group delay randomization in the higher frequency region.

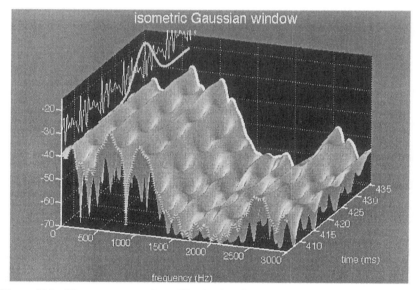

Figure 11.1. 3D spectrogram using a F0 adaptive Gaussian window. The vertical wall shows the original waveform and the windowing function.

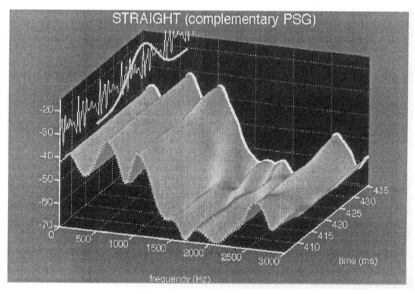

Figure 11.2. 3D spectrogram using a F0 adaptive complementary windows and spline based spectral smoothing. (This is the information representation in STRAIGHT.) The vertical wall shows the original waveform and the windowing function.

However, previous studies presented no dependable methodologies to extract control parameters of this group delay randomization from the speech signal under study. Therefore, this is the weakest part of the STRAIGHT system. We have developed new procedures to extend the source information extractor of the STRAIGHT system. The procedures also provide useful means to visualize voice–quality–related source parameters (Kawahara *et al.,* 1999b, 2000, 2001a).

2.2   F0 estimation

Speech signals are not exactly periodic: F0s and waveforms are always changing and fluctuating. The instantaneous frequency–based F0 extraction method used in STRAIGHT was proposed (Kawahara *et al.,* 1999b) to represent these nonstationary speech behaviors and was designed to produce continuous and high–resolution F0 trajectories suitable for high–quality speech modifications. Estimation of the aperiodicity measures in the frequency domain is dependent on this initial F0 estimate, which is itself based on a fixed–point analysis of a mapping from filter–center frequencies to their output instantaneous frequencies.

The F0 estimation method of STRAIGHT assumes that the signal has the following nearly harmonic structure, ,

$$x(t) = \sum_{k=1}^{N} a_k(t) \cos\left(\int_0^t \left(k\omega_o(\tau) + \omega_k(\tau)\right) d\tau + \phi_k \right) \tag{4}$$

where $a_k(t)$ represents a slowly changing instantaneous amplitude and $\omega_k(\tau)$ also represents slowly changing perturbation of the k-th harmonic component. In this representation, F0 is the instantaneous frequency of the fundamental component where $k=1$. The F0 extraction procedure also employs instantaneous frequencies of other harmonic components to refine F0 estimates.

By using band–pass filters with complex number impulse responses, filter–center frequencies and instantaneous frequencies of filter outputs provide an interesting means of extracting the sinusoidal components. Let $\lambda(\omega_c,t)$ be the mapping from the filter–center angular frequency ω_c to the instantaneous frequency of filter output. Then, angular frequencies of sinusoidal components are extracted as a set of fixed points   based on the following definition.

$$\text{)} = \left\{ \psi \mid \lambda(\psi,t) = \psi, \quad -1 \le \frac{\partial \lambda(\psi,t)}{\partial \psi} \le 1 \right\} \tag{5}$$

This relation between filter–center frequencies and harmonic components was reported by number of authors (Charpentier 1986, Abe *et al.*, 1995). A similar relation to resonant frequencies was also described in modeling auditory perception (Cooke 1993). In addition to these findings, geometrical properties of the mapping around fixed points were found to be very useful in analyzing source information (Kawahara *et al.*, 1999b).

The signal–to–noise ratio of the sinusoidal component and the background noise (represented as C/N: carrier-to-noise ratio hereafter) is approximately represented using first– and second–order partial derivatives of this mapping surface. Please refer to (Kawahara *et al.*, 1999b) for details. Combined with this C/N estimation method, the following nearly isotropic filter impulse response is designed.

$$w_s(t,\omega_c) = w(t,\omega_c) * h(t,\omega_c)\exp(j\omega_c t) \; , \tag{6}$$

$$w(t,\omega_c) = \exp\left(-\frac{\omega_c^2 t^2}{4\pi\eta^2}\right), \quad h(t,\omega_c) = \max\left\{0, \quad 1-\left|\frac{\omega_c t}{2\pi\eta}\right|\right\},$$

where * represents convolution and η represents a time stretching factor, which is slightly larger than 1 in order to refine frequency resolution. With a log–linear arrangement of filters, a fundamental harmonic component can be selected as the fixed point having the highest C/N. Then, the initial F0 estimate is used to select several (in our case, the lower three) harmonic components for refining the F0 estimate using C/N and the instantaneous frequency for each harmonic component.

Figure 11.3 shows an example to illustrate how the log–linear filter arrangement makes the fundamental component–related fixed point salient. It is clear that the mappings stay flat only around the fundamental component.

Figure 11.4 shows an example of STRAIGHT's source information display. It illustrates how C/N information is used for determining the fundamental component. In this figure, C/N information is shown on the top panel and the bottom panel. Please refer to the caption for further explanation.

As mentioned in the previous paragraph, this F0 estimation procedure comprises the C/N estimation for each filter output as its integral part, and it is applicable to aperiodicity evaluation (Kawahara *et al.*, 2001a).

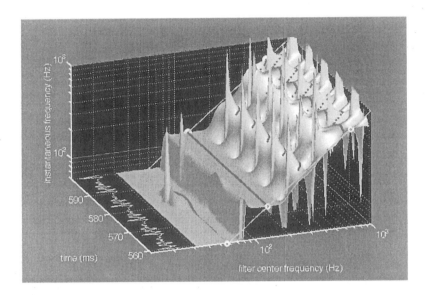

Figure 11.3. Three dimensional representation of the filter center frequency to the output instantaneous frequency map. The surface represents the mapping. Red dots represent the fixed point of the mapping. F0 can be found as fixed points on the unique flat surface. Note that other fixed points are not stable in time. Speech material is the sustained Japanese vowel /a/ spoken by a male speaker. Temporal stretch parameter = 1.1 was used. From Kawahara (2003b).

3   SYSTEMATIC DOWNGRADING

So far, the outline of a new versatile speech analysis, modification and synthesis system STRAIGHT has been introduced. STRAIGHT can resynthesize speech sounds from smoothed time–frequency representation, fundamental frequency and aperiodicity information at each time–frequency location. By using these three parameters, high–quality resynthesized speech, which is sometimes indistinguishable from natural speech in terms of naturalness, can be generated. These are objective parameters, and they can be controlled precisely in each physical dimension, which means that we can manipulate physical parameters of virtually natural speech. This section outlines how to use this versatility to investigate human performance in speech perception in natural, everyday situations. It is an exemplar–based approach; in other words, it is a type of data–mining or deductive approach. "Systematic downgrading" is one such strategy we have been adopting.

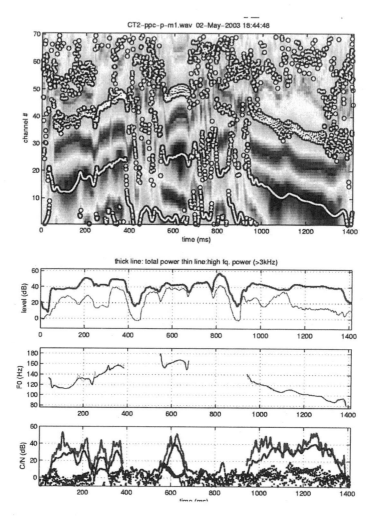

Figure 11.4. Extracted source information from the French singing by a male singer. The top panel represents fixed points extracted using a circle symbol. The overlaid image represents the C/N ratio for each filter channel (24 channels/octave covering from 80 Hz to 600 Hz). The lighter the color the higher the C/N. The middle panel shows the total energy (thick line) and the higher frequency (> 3 kHz) energy (thin line). The next panel illustrates an extracted F0. The bottom panel shows the C/N ratio for each fixed point. From Kawahara (2003b).

The strategy "systematic downgrading," was originally proposed in the context of our research on scat singing (Kawahara *et al.*, 2001b, 2002a, 2002b), where non-linguistic and para-linguistic information plays indispensable roles. The central idea of "systematic downgrading" is to keep test stimuli as ecologically relevant (in other words, highly natural) as possible.

The STRAIGHT–based morphing fulfills requirements for ecological relevance (high quality resynthesized speech) and precise control of physical parameters simultaneously.

The following steps outline systematic downgrading in the case of investigating regularities in voice quality.

10. Prepare the reference speech and the target speech that have typical voice quality.

11. Morph the reference speech to the target speech by careful manual transformation of parameters.

12. Extract regularities from the manual transformation and design series of approximation functions of the transformation.

13. Morph the reference speech by the approximation functions AND refine it with additional manual modifications.

14. Repeat steps (3) and (4) until a satisfactory approximation function is designed.

The procedure is a generalized version of the "null point procedure," which is a common practice to minimize disturbances to the system under study. It ensures that the critical subjective evaluation is performed only for high–quality (ecologically relevant) stimuli, which is especially important when conducting research on para– and nonlinguistic aspect of speech, because they are the most fragile attributes against various types of distortion. It should also be noted that step (3) is inevitably exploratory, even though multivariate analysis may help in the acquisition of insights. The first step of such an exploratory investigation would be to implement selective morphing. Similar examples can be found in our papers (Kawahara and Matsui, 2003, Matsui and Kawahara, 2003) and in a literature on emotional speech (Bulut *et al.*, 2002).

4 AUDITORY MORPHING

Morphing is a procedure to regenerate a signal from a representation on the shortest trajectory between anchor points in an abstract distance space with a distance metric d_{fx}. It is necessary to introduce an approximation that yields practical implementation of this general morphing procedure. One such approximation for speech morphing is to define the new distance d_{cp} as a composite operation of a coordinate transformation T and a localized distance metric d_{pp},

$$d_{fx} \cong d_{cp} = d_{pp}\left(S_{ref}(\omega, \tau), S_{tgt}\left(T(\omega, \tau)\right)\right),$$

(8)

where subscripts *ref* and *tgt* represent "reference" and "target," respectively. If the transformation *T* does not have any penalty due to the transformation, and if the localized distance metric is Euclidean, the morphing procedure is reduced to a linear interpolation on the reference coordinate. The procedure used in our morphing study is based on this approximation. This procedure is analogous to visual morphing when the time–frequency coordinate and the attributes on the coordinate system are replaced by the shape and color (including intensity and texture).

There are several technical issues involved in implementing the procedure. Specifically, the coordinate system and the localized distance metric must reflect auditory perceptual characteristics, and the transformation must be as simple as possible. In this article, the time–frequency plane is used as the coordinate system. The transformation is represented as a simple piecewise bilinear transformation, because, unlike the image morphing, the time–frequency coordinate is not isotropic. In our preliminary experiences, it was found that for morphing emotional speech samples digitized at 44.1 kHz and 16 bit, only up to five anchor points on a frequency axis at one temporal location and up to four temporal anchor points for one CV syllable were sufficient. For the fundamental frequency, it is relevant to morph the parameter in the log–frequency domain, because the F0 dynamics is represented in terms of a linear dynamical equation in the log–frequency domain (Fujisaki 1998). For the spectral density, morphing is calculated on dB representation, because it is one of relevant approximations of intensity perception. The time–frequency periodicity index (Kawahara *et al.*, 1999b, 2001a) is also transformed by the same mapping function. This is one of the contributing factors on speech segregation.

4.1   Illustration of the method

The following figures illustrate how to define and to use anchor points for auditory morphing based on STRAIGHT. Figure 11.5 shows anchor points that were manually defined for morphing two speech samples of a word spoken under different emotional contexts. The information given by Fig. 11.5 is used to deform the time–frequency coordinate system of the target speech to align it with the reference speech. Figure 11.6 shows how the target uniform time–frequency grid is deformed. By using this deformation, all spectral and source parameters in two end points (two speech examples) are aligned in the targets time–frequency coordinate system. This alignment simplifies morphing interpolation at each time–frequency location.

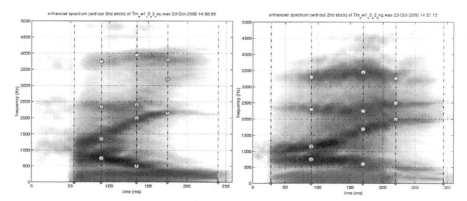

Figure 11.5. Smooth spectrographic representations of words played by a male actor under neutral (left) and angry (right) emotional conditions. Anchor points in the time–frequency domain are plotted as open circles and temporal anchors are plotted as vertical dash–dot lines. From Kawahara and Matsui (2003), © 2003 IEEE.

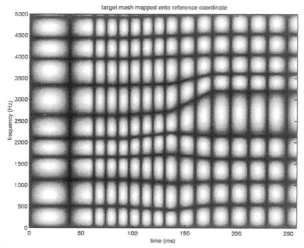

Figure 11.6. Regular time–frequency grid in the target coordinate system. It was transformed into the reference coordinate system. From Kawahara and Matsui (2003), © 2003 IEEE.

4.2 Quality of morphed speech

Figure 11.7 shows an example of naturalness evaluation results using ten subjects (six male and four female) on morphed emotional speech samples including the original natural speech samples (Matsui and Kawahara, 2003). Through this evaluation it was deduced that interpolated morphing, which includes simple analysis and synthesis by the STRAIGHT–based morphing procedure, provides manipulated speech samples indistinguishable from original natural speech samples in terms of naturalness. It may be safe to conclude that at least for naive subjects our morphing procedure is good enough to start systematic downgrading. However, for highly trained listeners, STRAIGHT is still found to add some specific coloring to processed speech, suggesting that there is still room for further investigations on STRAIGHT, and it is our responsibility to improve it.

5   EXTENSION AND APPLICATIONS

A new formulation of the STRAIGHT algorithm provides a means of traversing apparently different speech coding schema (Kawahara *et al.*, 2004). This extension will make it possible to test human performance in speech segregation under more realistic conditions. Application of STRAIGHT to an auditory inversion problem based on the AIM (Auditory Image Model, Patterson *et al.*, 1995) system yielded a unique procedure for auditory segregation (Irino *et al.*, 2003). Furthermore, it also provides a means to test the attractive hypothesis that human auditory system is designed to be capable of segregating information about the size and the shape of the vocal tract (Irino and Patterson, 2002).

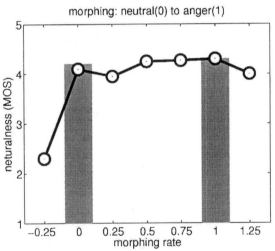

Figure 11.7. Naturalness evaluation of morphed and original speech samples. Bars represent original speech samples and circles represent morphed samples. The morphing rate "0" corresponds to "neutral" emotion and "1" corresponds to "anger" emotion. A Japanese word /koNnitiwa/ (hello in English) spoken by a male actor was used.

6   CONCLUSION

Testing human performance using ecologically relevant stimuli is crucial. STRAIGHT provide powerful means and strategies for doing this. This article outlined the underlying principles of STRAIGHT and the morphing procedure to provide general understanding for potential users of a new research strategy, "systematic downgrading." The strategy seems to open up new research possibilities of testing human performance without disturbing their natural conditions.

7   ACKNOWLEDGEMENT

This work is partially supported by an e-society project by the Japanese Ministry of Education and funding by Wakayama University. The first author is also an invited researcher of ATR, where STRAIGHT was invented.

References

Abe, T., Kobayashi, T., and Imai, S., 1995, Harmonics estimation based on instantaneous frequency and its application to pitch determination, *IEICE Trans. Information and Systems*, **E78-D(9)**:1188–1194.

Bulut, M., Narayanan, S.S., and Syrdal, A.K., 2002, Expressive speech synthesis using a concatenative synthesizer, *Proc. ICSLP'02, Denver*, **1**:1265–1268.

Charpentier, F.J., 1986, Pitch detection using the short–term phase spectrum, *Proc. ICASSP'86*, pp.113–116.

Cooke, M., 1993, *Modeling Auditory Processing and Organization*, Cambridge University Press, Cambridge, UK.

Fujisaki, H., 1998, A note on the physiological and physical basis for the phrase and accent components in the voice fundamental frequency contour, in: *Vocal Fold Physiology: Voice Production, Mechanisms and Functions*, O. Fujimura, ed., Raven Press, New York, pp.347–355.

Irino, T. and Patterson, R.D., 2002, Segregating information about the size and shape of the vocal tract using a time–domain auditory model: The stabilized wavelet Mellin transform, *Speech Communication*, **36(3–4)**:181–203.

Irino, T., Patterson, R.D., and Kawahara, H., 2003, Speech segregation based on fundamental event information using an auditory vocoder, *Proc. Eurospeech'03, Geneva*, pp.553–556.

Kawahara, H., Masuda-Katsuse, I., and de Cheveigné, A., 1999a, Restructuring speech representations using a pitch–adaptive time–frequency smoothing and an instantaneous–frequency–based F0 extraction, *Speech Communication*, **27(3–4)**:187–207.

Kawahara, H., Katayose, H., de Cheveigné, A., and Patterson, R.D., 1999b, Fixed–point analysis of frequency to instantaneous frequency mapping for accurate estimation of F0 and periodicity, *Proc. Eurospeech'99*, **6**:2781–2784.

Kawahara, H., Atake, Y., and Zolfaghari, P., 2000, Accurate vocal event detection method based on a fixed–point analysis of mapping from time to weighted average group delay, *Proc. ICSLP'2000*, pp.664–667.

Kawahara, H., Estill, J., and Fujimura, O., 2001a, Aperiodicity extraction and control using mixed mode excitation and group delay manipulation for a high quality speech analysis, modification and synthesis system STRAIGHT, *Proc. 2nd MAVEBA*, [CD ROM].

Kawahara, H. and Katayose, H., 2001b, Scat singing generation using a versatile speech manipulation system, STRAIGHT, *141st meeting of the Acoust. Soc. Amer., Chicago*, **109**:2425–2426.

Kawahara, H. and Katayose, H., 2002a, Scat generation research program based on STRAIGHT, a high–quality speech analysis, modification and synthesis system, *J. of IPSJ*, **43(2)**:208–218, [in Japanese].

Kawahara, H., 2002b, Systematic downgrading for investigating "naturalness" in synthesized singing using STRAIGHT: A high quality VOCODER, *143rd meeting of the Acoust. Soc. Amer., Pittsburgh*, p. 2334.

Kawahara, H. and Matsui, H., 2003, Auditory morphing based on an elastic perceptual distance metric in an interference–free time–frequency representation, *Proc. ICASSP'2003*, 1:256–259.

Kawahara, H., 2003b, Exemplar-based voice quality analysis and control using a high quality auditory morphing procedure based on STRAIGHT, *Proc. VOQUAL'03, Geneva*, pp.109-114

Kawahara, H., Banno, H., Irino, T., and Zolfaghari, P., 2004, Algorithm amalgam: morphing waveform based methods, sinusoidal models and STRAIGHT, *Proc. ICASSP'04* [accepted for publication].

Matsui, H. and Kawahara, H., 2003, Investigation of emotionally morphed speech perception and its structure using a high quality speech manipulation system, *Proc. Eurospeech'03, Geneva*, pp.2113–2116.

Patterson, R.D., Allerhand, M., and Giguere, C., 1995, Time–domain modeling of peripheral auditory processing: a modular architecture and a software platform, *J. Acoust. Soc. Amer.*, **98**:1890–1894.

Chapter 12

On Ideal Binary Mask As the Computational Goal of Auditory Scene Analysis

DeLiang Wang

Department of Computer Science and Engineering and Center of Cognitive Science, The Ohio State University, Columbus, OH
dwang@cis.ohio-state.edu

1   INTRODUCTION

In a natural environment, a target sound, such as speech, is usually mixed with acoustic interference. A sound separation system that removes or attenuates acoustic interference has many important applications, such as automatic speech recognition (ASR) and speaker identification in real acoustic environments, audio information retrieval, sound-based human computer interaction, and intelligent hearing aids design.

Because of its importance, the sound separation problem has been extensively studied in signal processing and related fields. Three main approaches are speech enhancement (Lim, 1983; O'Shaughnessy, 2000), spatial filtering with a microphone array (van Veen and Buckley, 1988; Krim and Viberg, 1996), and blind source separation using independent component analysis (ICA) (Lee, 1998; Hyvärinen *et al.*, 2001). Speech enhancement typically assumes certain prior knowledge of interference; for example, the standard spectral subtraction technique is easy to apply and works well when the background noise is stationary. However, the enhancement approach has difficulty in dealing with the unpredictable nature of general environments where a variety of intrusions, including nonstationary ones such as competing talkers, may occur. The objective of spatial filtering, or beamforming, is to estimate the signal that arrives from a specific direction through proper array configuration, hence filtering out interfering signals from other directions. With a large array spatial filtering can produce high-fidelity separation, and at the same time attenuate much signal reverberation. A main limitation of spatial filtering is what I call *configuration stationarity*: It has trouble tracking a target that moves around or switches between different sound sources. Closely related to spatial filtering is ICA-based blind source separation, which assumes statistical independence of sound sources and formulates the separation problem as that of estimating a demixing matrix. To make standard ICA formulation work requires a number of assumptions on the mixing process and the number of microphones (van der Kouwe *et al.*, 2001). ICA gives impressive separation results when its assumptions are met. On the other hand, the assumptions also limit the scope of the applicability. For example, the stationarity assumption on the mixing process – similar to the configuration stationarity in spatial filtering – is hard to satisfy when speakers turn their heads or move around.

While machine separation remains a challenge, the auditory system shows a remarkable capacity for sound separation, even monaurally (i.e. with one microphone). According to Bregman (1990), the auditory system organizes the acoustic input into perceptual streams, corresponding to different sources, in a process called auditory scene analysis (ASA). Bregman further asserts that ASA takes place in two stages in the auditory system: The first stage decomposes the acoustic mixture into a collection of sensory elements or segments, and the second stage selectively groups segments into streams. This two-stage conception corresponds in essence to an analysis-synthesis strategy. Major ASA cues include proximity in frequency and time, harmonicity, smooth transition, onset synchrony, common location, common amplitude and frequency modulation, and prior knowledge.

Research in ASA has inspired a series of computational studies to model auditory scene analysis (Weintraub, 1985; Cooke, 1993; Brown and Cooke, 1994; Ellis, 1996; Wang and Brown, 1999). Mirroring the above two-stage conception, computational auditory scene analysis (CASA) generally approaches sound separation in two main stages: segmentation and grouping. In segmentation, the acoustic input is decomposed into sensory segments, each of which likely originates from a single source, by analyzing harmonicity, onset, frequency transition, and amplitude modulation. In grouping, the segments that likely originate from the same source are grouped, based mostly on periodicity analysis. In comparison with other separation approaches, the main CASA success has been in monaural separation with minimal assumptions.[1] It also creates a new set of challenges and demands, such as reliable multipitch tracking and special handling of unvoiced speech.

In comparison with other well-established separation approaches, CASA faces a somewhat distinct issue: there is no consensus on how to quantitatively evaluate a CASA system (Rosenthal and Okuno, 1998). Almost every study adopts its own evaluation criteria. This is partly due to the fact that CASA is still in its infancy, but it may reflect deeper confusion on the computational goal of auditory scene analysis. The lack of common evaluation

---

[1.]More accurately, CASA also makes a number of assumptions, but such assumptions tend to conform to the constraints under which the auditory system operates.

criteria makes it difficult to document and communicate the progress made in the field. Sensible evaluation criteria can also serve as the guiding principle for model development.

This chapter intends to examine the goal of CASA. After analyzing the advantages and disadvantages of different computational objectives, I suggest ideal time-frequency (T-F) mask as the computational goal of auditory scene analysis. The remainder of the chapter is organized as follows. The next section reviews different CASA evaluation criteria. Section 3 is devoted to a general discussion of the CASA goal, including an analysis of several alternative CASA objectives. Section 4 introduces the ideal binary mask, analyzes their properties, and argues for their use as the CASA goal. Section 5 describes two models that explicitly estimate the ideal binary mask. Finally, Section 6 concludes the chapter.

2   CASA EVALUATION CRITERIA

CASA criteria that have been suggested can be divided into the following four categories: Direct comparisons between segregated target and premixing target, changes in automatic speech recognition (ASR) score, evaluation with human listening, and fit with biological data. Each is described below.

- **Comparison with premixing target**. Obviously this assumes the availability of premixing sound sources, which is not an unreasonable assumption for system evaluation. The evaluation criterion employed in Cooke's study (1993) is the match between a model-generated group of target elements and the group of elements in clean target speech. Brown and Cooke (1994) use a segregated target stream, which is a binary T-F mask, to resynthesize target speech and noise intrusion, and then calculate a normalized ratio between resynthesized speech and resynthesized noise. Subsequently, Wang and Brown (1999) use conventional signal-to-noise ratio (SNR), measured in decibels, between resynthesized speech and resynthesized noise. More tailored for speech, Nakatani and Okuno (1999) calculates spectral distortion by comparing the short-term spectra of segregated speech and those of clean speech. Bodden (1993) in his binaural model of speech segregation estimates a time-varying Wiener filter for each sound mixture, which consists of energy ratios between the target speech and the mixture within critical bands.

- **ASR measure**. A main motivation behind research on speech separation is to improve ASR performance in the presence of acoustic interference. So it is natural to evaluate a CASA model in terms of ASR score. This measure is used in the Weintraub model (1985) – probably the earliest CASA study. The evaluation metric is straightforward: Measuring changes in the recognition score using a standard ASR system before and after sound mixtures are segregated by the CASA model in question. Early ASR evaluations produce ambiguous results, partly because processing stages in CASA tend to distort the target signal, creating a mismatch between segregated signal and clean signal used ASR training. More recent attempts have yielded better outcomes (see, for example, Glotin, 2001).

■ **Human listening**. Human listeners can be involved in evaluating a computational model in terms of speech intelligibility on original mixtures and on segregated speech (1988; 1990). An improvement in speech intelligibility would lend support to the value of the model. However, human listeners are very good at segregating a sound mixture, and this creates a potential confound for using listeners to test model output. One practical difficulty is that in order to give room for a model to improve on intelligibility, interference must be very strong, which could be exceedingly hard for models to perform. Listeners with hearing loss may be better suited for such evaluation as it is well-known that people with sensorineural impairment have greater difficulty in segregating target speech in a noisy environment (Moore, 1998). Of course, if the objective of the model is to improve hearing of abnormal listeners or that of normal listeners in highly noisy environments, this evaluation methodology is the best choice. Ellis (1996) made a different use of human listening in evaluation: His listeners were used to score the resemblance of segregated sounds to component sounds in the mixture.

■ **Fit with biological data**. Some CASA researchers are interested in modeling the human ASA process, while some others are interested in elucidating neurobiological mechanisms underlying ASA. For such models, the main evaluation criterion is how well the models account for known perceptual or neurobiological data. Wang (1996) sought to model a number of ASA phenomena on the basis of a neural oscillator network (see also Norris, 2003). McCabe and Denham (1997) proposed a different neural network that simulates psychophysical results on auditory streaming. Recently, Wrigley and Brown (2004) put forward a neural oscillator model of auditory attention and used it to quantitatively simulate a set of psychological data.

3   WHAT IS THE GOAL OF CASA?

Different evaluation criteria tend to reflect different goals of computational models, whether or not they are explicitly laid out. This raises the question of what should be the goal of CASA? This is a very important question since its answers bear directly on the research agenda and determine whether computational efforts lead to real progress towards ultimately solving the CASA problem.

To address this question, it might be helpful to put CASA in a broad context of perception since CASA purports to model auditory scene analysis, which is a major process of auditory perception. So a larger question is, what is the goal of perception? This question, raised in the most general form, would fall under the realm of philosophy, and indeed philosophers have debated this issue for centuries. What we are concerned here is the information processing perspective, which is shared by human and machine perception. From this perspective, Gibson (1966) considers perceptual systems as ways of seeking and extracting information about the environment from the sensory input. In the visual domain, Marr (1982) states that the purpose of vision is to produce a visual description of the environment for the viewer. By extrapolating Marr's statement to the auditory domain, the purpose of audition would be to produce an auditory description of the environment for the listener. It is worth noting, according to the above views, that perception is a process private to the perceiver despite the fact that the physical environment is common to different perceivers.

According to Bregman (1990), the goal of ASA is to produce separate auditory streams from sound mixtures, each stream corresponding to an acoustic event. This would imply that the goal of CASA is to computationally extract individual streams from sound mixtures. To make this description more meaningful, however, further constraints need to be observed:

- To qualify as a stream a sound must be audible on its own. In other words, the intensity of the sound at the eardrum must exceed a certain sound level, referred to as the absolute threshold (Moore, 2003).

- The number of streams that can be segregated at a time must be limited. This limit is directly related to the capacity of auditory attention. In a comprehensive account, Cowan recently concluded that the capacity of attention is about four (Cowan, 2001). This implies that the auditory system cannot segregate more than 4 streams simultaneously. While a listener may be able to segregate up to 4 tones or steady vowels, in a very noisy environment such as a cocktail party, the attentional capacity may reduce to figure-ground separation, i.e.

attending to only a foreground stream with a general awareness of the background.

- A fundamental fact in auditory perception is auditory masking (Moore, 2003). Roughly speaking, auditory masking refers to the phenomenon that within a critical band a stronger signal tends to mask a weaker one. When a sound is masked, it is eliminated from perception as if the sound never reached the ear.

- ASA results depend on sound types. Say we listen to mixtures of two equally-loud sounds. If the sounds are two tones well separated in frequency or two speech utterances, we can readily segregate them. On the other hand, if the sounds are while noise and pink noise we are completely incapable of any segregation.

With the above analysis in mind, we now discuss some alternative CASA objectives. The first objective, which might be called the gold standard, is simply to segregate all sound sources from a sound mixture. If this standard could be reached, it would be the ideal goal of CASA, at least from an engineering standpoint. On the other hand, the goal is clearly beyond what the human listener can do; just observe for yourself how many conversations you can follow in a cocktail party. It is probably also an unrealistic computational goal if the system has just one or two microphones.

Another alternative objective is to enhance ASR. This objective has the advantage that it directly relates to one of the primary motivations for CASA research. The objective is also straightforward to evaluate as discussed in the last section. This objective has several drawbacks. One drawback is that it is narrowly focused on speech. Although speech is a vital type of acoustic signal for humans, it is by no means the only important signal to us. What about music, or other environmental sounds? For music in particular, it is hard to characterize music perception as a recognition process. A deeper issue with recognition as the goal is that perceiving is more than recognizing (Treisman, 1999). Perceiving has, in addition of recognition, all the current details of events, such as how they sound like, where they are, whether they are approaching or receding, and many other details about them. Such details are crucial for the perceiver to decide how to act. Also it is not clear how the ASR objective can account for the fact that new things unheard before can be perceived as well.

The third alternative is to enhance human listening. A main advantage of this objective is the close coupling with auditory perception. Also a primary motivation of studying CASA is to improve hearing prosthesis for listeners with hearing impairment as well as hearing of normal listeners in very noisy environments. However, this objective is specifically tailored to human listening and there are other applications that do not directly involve humans, such as audio information retrieval. There are also practical difficulties for computational researchers in terms of required expertise for conducting human experiments.

These alternative objectives have their advantages and disadvantages. A desirable objective should be generally consistent with the above analysis on human auditory scene analysis, and be comprehensive enough to apply to different types of acoustic signal and different application domains. The objective should not consider just recognition performance or human listening, but at the same time it should be consistent with such criteria. The simplicity of the objective and easiness to apply are also desirable so that a researcher need not wait for a long time to find out how well a provisional model works. In the next section, I present the ideal time-frequency mask as a putative goal of CASA.

4   IDEAL BINARY MASK AS THE GOAL OF CASA

As discussed in Section 3, the gold-standard objective is probably unrealistic. A more realistic objective is to segregate a target signal from the mixture. Then the objective becomes that of figure-ground separation. This begs the question of what should be regarded as the target? Generally speaking, what the target is depends on external input as well as intention; it is closely related to the study of attention, in particular what attracts attention (Pashler, 1998). From a practical standpoint, what constitutes the target is task-dependent and often unambiguous. For the purpose of our discussion, we assume that the target is known. We also assume, for the sake of evaluation, the availability of premixing target signal and interference.

A widely accepted representation in CASA is the two-dimensional time-frequency representation where the time dimension consists of a sequence of time frames and the frequency dimension consists of a bank of auditory filters (e.g. gammatone filters). This representation is consistent with accounts of human ASA and auditory physiology. Within this representation, the key consideration behind the notion of the ideal binary mask is to retain the time-frequency regions of a target sound that are stronger than the interference, and discard the regions that are weaker than the interference. More specifically, an ideal mask is a binary matrix, where 1 indicates that the target energy is stronger than the interference energy within the corresponding T-F unit and 0 indicates otherwise. This definition implies a 0-dB SNR criterion for mask generation, and other SNR criteria are possible too (see below). Figure 12.1 illustrates the ideal mask for a mixture of a male utterance and a female utterance, where the male utterance is regarded as target. The overall SNR of the

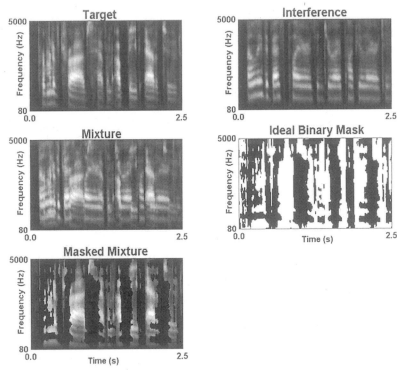

Figure 12.1. Illustration of the ideal binary mask. **Top left**: Two-dimensional T-F representation of the target utterance ("Primitive tribes have an upbeat attitude"). The figure displays the rectified responses of the gammatone filterbank with 128 channels. **Top right**: Corresponding representation of the interfering utterance ("Only the best players enjoy popularity"). **Middle left**: Corresponding representation of the mixture. **Middle right**: Ideal T-F binary mask, where white pixels indicate 1 and black pixels 0. **Bottom left**: Masked mixture using the ideal binary mask.

mixture is 0 dB. The top left panel of Figure 12.1 shows the T-F representation of the target utterance, the top right panel the representation of the interfering utterance, and the middle left panel the representation of the mixture. For this mixture, the ideal mask is shown in the middle right panel. The bottom left panel of the figure shows the result of ideal masking on the mixture. Compared with the original mixture, the masked mixture is much closer to the clean target. Listening to the masked mixture one can clearly hear the target utterance while no trace of interference is audible.

Binary masks have been used as an output representation in the CASA literature (Brown and Cooke, 1994; Wang and Brown, 1999). Related to binary masks is the observation that different speech utterances tend to be orthogonal in a high-resolution time-frequency representation because the energy of a single utterance tends to be sparsely distributed (Jourjine *et al.*, 2000; Roweis, 2001). This observation obviously does not hold when an acoustic background is babble noise or contains broadband intrusions. To my knowledge, the papers of Hu and Wang (2001) and Roman *et al.* (2001) are the earliest studies that suggest the use of the ideal binary mask (see also Roman *et al.*, 2003; Hu and Wang, 2004). Note that the definition of the ideal binary mask does not assume orthogonality among sound sources.

The ideal binary mask has a number of desirable properties:

- Flexibility. With the same mixture, the definition leads to different masks depending on what the target is. It is consistent with the perceptual observation that the same environment can be perceived in different ways by different perceivers.

- Well-definedness. The ideal mask is well defined no matter how many intrusions are in the scene. One may also identify multiple targets from the same mixture, with multiple processors that have different target definitions.

- The ideal binary mask sets the ceiling performance for all binary masks.

- The ideal mask is broadly consistent with ASA constraints in terms of audibility and segregation capacity. In particular, it has direct correspondence with the auditory masking phenomenon

When a gammatone filterbank is used for generating the time-frequency representation, a technique introduced by Weintraub (1985) can be used to resynthesize a waveform signal from a binary mask (see also Brown and Cooke, 1994; Wang and Brown, 1999). One can then conduct listening tests on resynthesized signal. The ideal binary mask produces high quality resynthesized target unless the mixture SNR is very low.

Recent research on missing-data speech recognition provides an effective bridge between a segregated mask and ASR (Cooke *et al.*, 2001). The main idea of missing-data recognition is to adapt the standard HMM recognizer so that recognition decisions are based only on reliable T-F units while marginalizing unreliable or missing T-F units. Cooke *et al.* (2001) found that the *a priori* mask – defined according to whether the mixture energy is within 3 dB of the target energy – used in conjunction missing-data recognition yields excellent recognition performance. Similar performance is obtained by Roman *et al.* (2003) using the ideal binary mask. Moreover, the study of Roman *et al.* (2003) found that deviations from the ideal binary mask lead to gradual degradation in speech recognition performance.

The ideal binary mask has been recently tested in human speech intelligibility experiments. As noted earlier, the definition of the ideal mask uses the 0-dB SNR criterion within individual T-F units. However, one can produce different ideal masks using different local SNR criteria. Brungart *et al.* (in preparation) tested a range of local SNR criteria around 0 dB using ideal masking on speech mixtures involving one target talker and 1 to 3 competing talkers. All talkers have equal overall loudness, or the SNR between the target and a single competing talker is zero. Their experiments showed that, within the local SNR range from -5 dB to 5 dB, ideal masking produces intelligibility scores near 100% in all mixtures involving 2, 3, and 4 talkers. In addition, the intelligibility score decreases systematically towards higher or lower SNR criteria. Note that for a fixed mixture a very high SNR criterion leads to a mask with very few 1's, hence very little target energy; a very low SNR criterion leads to a mask close to an all-1 mask, hence very little segregation. Their results also show that, for mixtures with very low SNR, ideal masking improve speech intelligibility dramatically (see also Roman *et al.*, 2001).

Finally an analogy may be drawn between auditory binary masking and visual occlusion. Figure 12.2 illustrates occlusion with a natural image of water lilies, where a lily in the front occludes the objects in the back. Visual occlusion may be considered as an instance of binary masking, in which the pixels of a front surface are assgined 1 in the mask and those of the occluded surfaces are assigned 0. Moreover, when an observer attends to a particular object in an image (say the lily near the center of Figure 12.2), this process of attending is analogous to ideal binary masking where the pixels of the attended object correspond to 1's in the mask and the remaining pixels correspond to 0's.

Figure 12.2. A natural image of water lilies.

5 ESTIMATION OF THE IDEAL BINARY MASK

The ideal binary mask clearly quantifies the computational goal of CASA. Guided by this goal, we have made conscious effort to compute the ideal mask. This section describes two models that explicitly estimate the ideal binary mask.

5.1 Monaural segregation of voiced speech

Voiced speech segregation has been a primary topic in CASA. For voiced speech, harmonicity is the essential cue for segregation. Earlier CASA models can segregate much of the low-frequency energy, but have trouble segregating high-frequency components. It is well-known that the auditory system can resolve the first few harmonics, while higher harmonics are unresolved. Psychoacoustic research suggests that the auditory system may use different mechanisms to deal with resolved and unresolved harmonics (Carlyon and Shackleton, 1994; Bird and Darwin, 1997). Subsequently, Hu and Wang (2003; 2004) developed a CASA model that employs different mechanisms in the low- and the high-frequency range. The model follows the general two-stage processing (see Section 1): Segmentation and grouping. Building on the output from the Wang-Brown model (1999) that works well in the low-frequency range, Hu and Wang proposed a psychoacoustically motivated method for tracking target pitch contours.

With the results of target pitch tracking, the model then labels individual T-F units. In the low-frequency range, a T-F unit is labeled by comparing its response periodicity and the extracted target pitch period. In the high-frequency range, wide bandwidths of auditory filters cause the filters to respond to multiple unresolved harmonics of voiced speech. These responses are amplitude modulated due to beats and combinational tones (Helmholtz, 1863). Furthermore, response envelopes fluctuate at the frequency that corresponds to the fundamental frequency of speech. Hence, the model labels a high-frequency unit by comparing its amplitude modulation (AM) rate with the extracted pitch frequency. To derive AM rates Hu and Wang have employed a sinusoidal modeling technique; specifically, a single sinusoid is used to model AM within a certain range of target pitch and the derivation of AM rates can then be formulated as an optimization problem. With appropriately chosen initial values, the optimization problem can be solved efficiently using an iterative gradient descent technique. With labeled T-F units, the model generates segments in the low-frequency range based on temporal continuity and cross-channel correlation between responses of adjacent frequency channels, and in the high-frequency range based on temporal continuity and common AM among adjacent filter responses. Segments thus formed then expand iteratively, and the resulting collection of the segments with the target label gives the segregated target which is represented by a binary T-F mask.

Figure 12.3 illustrates the result of ideal mask estimation for voiced speech segregation. The top left panel of the figure shows the T-F representation of a voiced utterance which is the target. The top right panel shows the mixture of the utterance with a 'cocktail party' noise from Cooke (1993). The middle left panel shows the ideal binary mask for the mixture, and the middle right panel the estimated mask. The estimated mask is reasonably close to the ideal one. The bottom left panel gives the result of ideal masking on the mixture, and the bottom right panel the result of masking using the estimated mask.

The model of Hu and Wang (2004) produces substantially better performance than previous models, especially in the high-frequency range. In terms of systematic SNR evaluation, one may treat the resynthesized signal from the ideal binary mask as signal because the ideal mask represents the

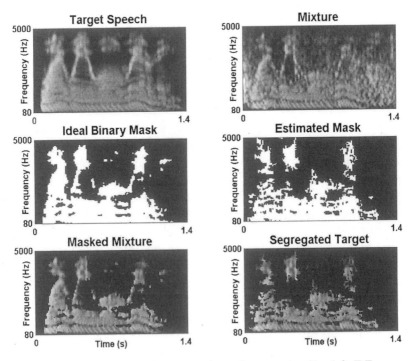

Figure 12.3. Ideal mask estimation for monaural speech segregation. **Top left**: T-F representation of the target utterance ("Why were you all weary"). **Top right**: T-F representation of the mixture of the target and the cocktail party noise. **Middle left**: Ideal binary mask for the mixture. **Middle right**: Estimated binary mask for the mixture. **Bottom left**: Masked mixture using the ideal mask. **Bottom right**: Masked mixture using the estimated mask.

computational goal. The model then yields 5.2 dB improvement over the Wang and Brown model (1999), which had the representative performance of earlier CASA systems. It also has 6.4 dB gain over the standard spectral subtraction method in speech enhancement. Similar improvements are obtained with conventional SNR metric using premixing speech as signal.

5.2   Binaural speech segregation

It is well known that people can selectively attend to a single voice at a noisy cocktail party. Spatial location is believed to play an important role in cocktail party processing. How to simulate this perceptual ability, known as the cocktail-party problem (Cherry, 1953), is a great computational challenge.

Guided by the notion of the ideal binary mask, Roman *et al.* (2003) developed a new location-based approach to speech segregation. Their model uses the binaural cues of interaural time difference (ITD) and interaural intensity difference (IID) extracted from a KEMAR dummy head that realistically simulates the filtering process of the head, torso and external ear. They observe that, within a narrow frequency band, modifications to the relative energy of the target source to the interfering energy trigger systematic changes in the values of the binaural cues. For a given spatial configuration, this interaction produces characteristic clustering in the binaural feature space. Consequently, the model performs independent supervised learning for different spatial configurations and different frequency bands in the joint ITD-IID feature space. More specifically, they formulate the estimation of the ideal binary mask as a binary Bayesian classification problem, where the hypothesis is whether the target is stronger than the overall interference within a single T-F unit. Then a nonparametric method (kernel density estimation) is used to estimate likelihood functions in the ITD-IID space, which are then used in maximum *a posteriori* (MAP) decision making.

Figure 12.4 illustrates the result of estimating the ideal binary mask for natural speech segregation, using the same mixture shown in Figure 12.1. The top right panel shows the ideal binary mask, and the bottom right panel the estimated mask. The match between the two masks is excellent. Finally, the bottom left panel displays the result of masking the mixture using the estimated mask (cf. bottom left panel of Figure 12.1).

The resulting model was systematically evaluated in two-source and three-source configurations, and estimated binary masks approximate the ideal ones extremely well. In terms of conventional SNR evaluation, the model produces large and consistent SNR improvements over original mixtures. The SNR gains are as large as 13.8 dB in the two-source case and 11.3 dB in the three-source case. A comparison with the Bodden model (1993), which estimates a Wiener filter, shows that the Roman *et al.* model produces 3.5 dB

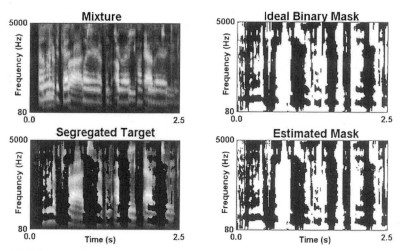

Figure 12.4. Ideal mask estimation for binaural speech segregation. **Top left**: the same mixture shown in Figure 12.1. **Top right**: Ideal binary mask for the mixture (also shown in Figure 12.1). **Bottom right**: Estimated binary mask. **Bottom left**: Masked mixture using the estimated mask.

improvement in the most favorable conditions for the Bodden model, and in other conditions the improvement is significantly greater. In addition to SNR evaluation, they performed an ASR evaluation by feeding estimated binary masks to a missing-data recognizer (Cooke *et al.*, 2001), and the model yields large ASR improvements compared to direct recognition of mixtures. Also, the model was evaluated on speech intelligibility with human listeners. Because people excel at ASA and achieve near perfect intelligibility unless interference is severe, the tests used three low SNR levels: 0 dB, -5 dB and -10 dB (measured at the better ear). The general finding is that the algorithm improves human intelligibility for the tested conditions, and the improvement becomes larger as the SNR decreases – as large as an increase from an intelligibility score of 20% to 80% at -10 dB.

6   CONCLUSION

In his famous treatise of computational vision, Marr (1982) makes a compelling argument for separating different levels of analysis in order to understand complex information processing. In particular, the computational theory level, concerned with the goal of computation and general processing strategy, must be separated from the algorithm level, or the separation of *what* from *how*. This chapter is an attempt at a computational-theory analysis of auditory scene analysis, where the main task is to understand the character of the CASA problem.

My analysis results in the proposal of the ideal binary mask as a main goal of CASA. This goal is consistent with characteristics of human auditory scene analysis. The goal is also consistent with more specific objectives such as enhancing ASR and speech intelligibility. The resulting evaluation metric has the properties of simplicity and generality, and is easy to apply when the pre-mixing target is available. The goal of the ideal binary mask has led to effective for speech separation algorithms that attempt to explicitly estimate such masks.

7   ACKNOWLEDGMENTS

The author thanks G. Hu and N. Roman for their assistance in figure preparation. This research was supported in part by an NSF grant (IIS-0081058) and an AFOSR grant (FA9550-04-1-0117).

References

Bird, J. and Darwin, C.J., 1997, Effects of a difference in fundamental frequency in separating two sentences, in: *Psychophysical and Physiological Advances in Hearing*, A.R. Palmer, *et al.*, ed., Whurr, London,

Bodden, M., 1993, Modeling human sound-source localization and the cocktail-party-effect, *Acta Acust.* **1**: 43-55.

Bregman, A.S., 1990, *Auditory Scene Analysis*, MIT Press, Cambridge MA.

Brown, G.J. and Cooke, M., 1994, Computational auditory scene analysis, *Computer Speech and Language* **8**: 297-336.

Brungart, D., Chang, P., Simpson, B., and Wang, D. L., in preparation.

Carlyon, R.P. and Shackleton, T.M., 1994, Comparing the fundamental frequencies of resolved and unresolved harmonics: Evidence for two pitch mechanisms?, *J. Acoust. Soc. Am.* **95**: 3541-3554.

Cherry, E.C., 1953, Some experiments on the recognition of speech, with one and with two ears, *J. Acoust. Soc. Am.* **25**: 975-979.

Cooke, M., 1993, *Modelling Auditory Processing and Organization*, Cambridge University Press, Cambridge U.K.

Cooke, M., Green, P., Josifovski, L., and Vizinho, A., 2001, Robust automatic speech recognition with missing and unreliable acoustic data, *Speech Comm.* **34**: 267-285.

Cowan, N., 2001, The magic number 4 in short-term memory: a reconsideration of mental storage capacity, *Behav. Brain Sci.* **24**: 87-185.

Ellis, D.P.W., 1996, *Prediction-driven computational auditory scene analysis*, Ph.D. Dissertation, MIT Department of Electrical Engineersing and Computer Science.

Gibson, J.J., 1966, *The Senses Considered as Perceptual Systems*, Greenwood Press, Westport CT.

Glotin, H., 2001, *Elaboration et étude comparative de systèmes adaptatifs multi-flux de reconnaissance robuste de la parole: incorporation d'indices de voisement et de localisation*, Ph.D. Dissertation, Institut National Polytechnique de Grenoble.

Helmholtz, H., 1863, *On the Sensation of Tone* (A.J. Ellis, Trans.), Dover Publishers, Second English ed., New York.

Hu, G. and Wang, D.L., 2001, Speech segregation based on pitch tracking and amplitude modulation, in *Proceedings of IEEE Workshop on Applications of Signal Processing to Audio and Acoustics*, pp. 79-82.

Hu, G. and Wang, D.L., 2003, Monaural speech separation, in: *Advances in Neural Information Processing Systems (NIPS'02)*, MIT Press, Cambridge MA, pp. 1221-1228.

Hu, G. and Wang, D.L., 2004, Monaural speech segregation based on pitch tracking and amplitude modulation, *IEEE Trans. Neural Net.*, in press.

Hyvärinen, A., Karhunen, J., and Oja, E., 2001, *Independent Component Analysis*, Wiley, New York.

Jourjine, A., Rickard, S., and Yilmaz, O., 2000, Blind separation of disjoint orthogonal signals: demixing N sources from 2 mixtures, in *Proceedings of IEEE ICASSP*, pp. 2985-2988.

Krim, H. and Viberg, M., 1996, Two decades of array signal processing research: The parametric approach, *IEEE Sig. Proc. Mag.* **13**: 67-94.

Lee, T.-W., 1998, *Independent Component Analysis: Theory and Applications*, Kluwer Academic, Boston.

Lim, J., ed., 1983, *Speech Enhancement*, Prentice Hall, Englewood Cliffs NJ.

Marr, D., 1982, *Vision*, Freeman, New York.

McCabe, S.L. and Denham, M.J., 1997, A model of auditory streaming, *J. Acoust. Soc. Am.* **101**: 1611-1621.

Moore, B.C.J., 1998, *Cochlear Hearing Loss*, Whurr Publishers, London.

Moore, B.C.J., 2003, *An Introduction to the Psychology of Hearing*, Academic Press, 5th ed., San Diego, CA.

Nakatani, T. and Okuno, H.G., 1999, Harmonic sound stream segregation using localization and its application to speech stream segregation, *Speech Comm.* **27**: 209-222.

Norris, M., 2003, *Assessment and extension of Wang's oscillatory model of auditory stream segregation*, Ph.D. Dissertation, University of Queensland School of Information Technology and Electrical Engineering.

O'Shaughnessy, D., 2000, *Speech Communications: Human and Machine*, IEEE Press, 2nd ed., Piscataway NJ.

Pashler, H.E., 1998, *The Psychology of Attention*, MIT Press, Cambridge MA.

Roman, N., Wang, D.L., and Brown, G.J., 2001, Speech segregation based on sound localization, in *Proceedings of IJCNN*, pp. 2861-2866.

Roman, N., Wang, D.L., and Brown, G.J., 2003, Speech segregation based on sound localization, *J. Acoust. Soc. Am.* **114**: 2236-2252.

Rosenthal, D.F. and Okuno, H.G., ed., 1998, *Computational Auditory Scene Analysis*, Lawrence Erlbaum, Mahwah NJ.

Roweis, S.T., 2001, One microphone source separation, in: *Advances in Neural Information Processing Systems (NIPS'00)*, MIT Press,

Stubbs, R.J. and Summerfield, Q., 1988, Evaluation of two voice-separation algorithms using normal-hearing and hearing-impaired listeners, *J. Acoust. Soc. Am.* **84**: 1236-1249.

Stubbs, R.J. and Summerfield, Q., 1990, Algorithms for separating the speech of interfering talkers: Evaluations with voiced sentences, and normal-hearing and hearing-impaired listeners, *J. Acoust. Soc. Am.* **87**: 359-372.

Treisman, A., 1999, Solutions to the binding problem: progress through controversy and convergence, *Neuron* **24**: 105-110.

van der Kouwe, A.J.W., Wang, D.L., and Brown, G.J., 2001, A comparison of auditory and blind separation techniques for speech segregation, *IEEE Trans. Speech Audio Process.* **9**: 189-195.

van Veen, B.D. and Buckley, K.M., April 1988, Beamforming: A versatile approach to spatial filtering, *IEEE ASSP Magazine*, pp. 4-24.

Wang, D.L., 1996, Primitive auditory segregation based on oscillatory correlation, *Cognit. Sci.* **20**: 409-456.

Wang, D.L. and Brown, G.J., 1999, Separation of speech from interfering sounds based on oscillatory correlation, *IEEE Trans. Neural Net.* **10**: 684-697.

Weintraub, M., 1985, *A theory and computational model of auditory monaural sound separation*, Ph.D. Dissertation, Stanford University Department of Electrical Engineering.

Wrigley, S.N. and Brown, G.J., 2004, A computational model of auditory selective attention, *IEEE Trans. Neural Net.*, in press.

Chapter 13

The History and Future of CASA

Malcolm Slaney
IBM Almaden Research Center
malcolm@ieee.org

1   INTRODUCTION

In this chapter I briefly review the history and the future of computational auditory scene analysis (CASA). Auditory scene analysis describes the process we use to understand the world around us. Our two ears hear a cacophony of sounds and understand that the periodic tic-toc comes from a clock, the singing voice comes from a radio and the steady hum is coming from the refrigerator.

The field of computational auditory scene analysis crystallized with the publication of Al Bregman's book "Auditory Scene Analysis" (1990). The commonly understood goal is to listen to a cacophony of sounds and separate the sounds from the mixture, just as humans do. I argue that this is *not* what people do. In this review, I will describe some progress to date towards modeling human sound separation, and review why this is the wrong direction for those of us interested in modeling human perception. Instead, we should be thinking about sound understanding. Auditory scene analysis and sound understanding are inextricably linked. Sound understanding is clearly a much harder problem, but should provide a better model of human sound separation abilities.

In particular, this chapter makes two related points. 1) We need to consider a richer model of sound processing in the brain, and 2) human sound separation work should not strive to generate acoustic waveforms of the separated signals. Towards this goal, this paper reviews the use of a correlogram for modeling perception and understanding sounds, the success at inverting the correlogram representation and turning it back into sound, and then summarizes recent work that questions the ability of humans to isolate separate representations of each sound object.

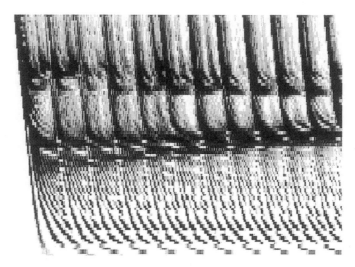

*Figure 13.1.*Simulated auditory nerve firings for the sound "Re". The vertical axis is arranged tonotopically, while time flows horizontally from left to right.

2   AUDITORY MODELING

This section reviews three aspects of auditory modeling: the cochleagram, the correlogram, and their inversion to recover the original sound. This inversion process is interesting, not only because it demonstrates the fidelity of the representation, but also because many sound-separation systems perform their separation in one of these two domains and then want to demonstrate their performance by resynthesizing the cleaned-up sound.

2.1   The Cochleagram

The cochlea transcribes the sounds into a stream of neural firings carried by the auditory nerve. These neural firings are often arranged in order of each nerve's best frequency to form a cochleagram. The richness of the cochlear data is shown in Figure 13.1. All sound that is perceived travels along the auditory nerve, so it is a complete representation of the perceived sound.

Figure 13.1 shows the output of a cochlear model for the sound "Re." Horizontal bands are at positions along the basilar membrane where there is significant spectral energy and correspond to the formants that are used to describe speech signals. More interestingly, there is a volley of firings at a regular interval represented by the (slanted) vertical lines. These periodic firings correspond to the energy imparted into the signal by the glottis and their interval directly corresponds to the glottal pitch.

2.2   The Correlogram

The richness and redundancy of the cochleagram suggests the need for an intermediate representation known as the correlogram. The correlogram was first proposed as a model of auditory perception by James Licklider (1951). His goal was to provide a unified model of pitch perception, but the correlogram was also been widely used as a model for the extraction and representation of temporal structure in sound.

The correlogram summarizes the information in the auditory firings using the auto-correlation of each cochlear channel (a channel, in this work, is defined as the firing probabilities of auditory nerves that innervate any one portion of the basilar membrane.) Its goal is collapse and summarize the repetitive temporal patterns shown in Figure 13.1.

The correlogram has met with great success in a number of areas. Meddis and his colleagues (Meddis and Hewitt 1991, Slaney and Lyon, 1990) have demonstrated the ability of a correlogram to model human pitch perception. Assman and Summerfield (1990) and others have shown that the correlogram is a useful representation when modeling the ability of humans to perceive two simultaneous spoken vowels. If anything, models using an ideal correlogram as their internal representations perform even better than humans.

The correlogram has served as a compelling visualization tool. In one auditory example created by Steve McAdams and Roger Reynolds, an oboe is split into even and odd harmonics. When the even and odd harmonics are played together at their original frequencies it sounds like the original oboe. But then independent frequency modulation (vibrato) is added to the even and the odd harmonics. The oboe separates into two sounds. The odd harmonics sound like a clarinet because a clarinet has mostly odd harmonic content, while the even harmonics go up an octave in pitch and sound like a soprano. A correlogram of this sound is quite striking: two sets of gray dots seem to float independently on the screen. Our task is simply a matter of identifying the dots with the common motion—using whatever tools make sense from a perceptual point of view—synthesizing two partial correlograms, and then resynthesizing the original sound.

2.3   Auditory Inversion

Given the grouping of the sound energy, a sound is synthesized from the selected neural representation, to allow human ears (and funding agencies) hear the separated sounds. Towards this goal, a number of us spent years

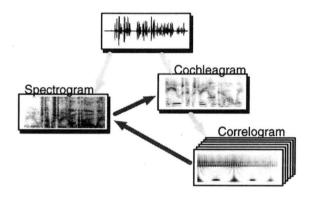

Figure 13.2. The correlogram inversion process. From waveform to correlogram and back.

developing the algorithms that allow us to turn a correlogram movie back into sound (Slaney 1996). The process requires two steps, as shown in Figure 13.2: 1) cochleagram inversion and 2) correlogram inversion.

From the cochlear representation (the information available on the auditory nerve, or a cochleagram) we invert the loudness compression (undo the automatic gain control in Lyon's model) run the auditory nerve probabilities backwards through the auditory filters, sum the backwards outputs into a single time-domain signal, and then repeat. This procedure, using an idealized cochlear model such as the one produced by Lyon can be done without perceptual differences between the original sound and the sound inversion from the cochleagram.

This procedure was used by Weintraub (1986) in the first real CASA system. He used the correlogram to track the pitch of two different sounds. Each channel was assigned to one sound object or the other, depending on which speaker's pitch was dominant. Then those cochleagram channels from each speaker were grouped, turned back into sound and applied to a speech recognizer. He realized a small improvement in speech recognition performance in a digit identification task.

The second, and more difficult problem, is inverting the correlogram to produce the original cochleagram. The important insight is to realize that each row of the correlogram is a time-varying autocorrelation. An autocorrelation contains the same information as the power spectrum (via an FFT) so each row of the correlogram can be converted into a spectrogram. Spectrogram inversion can be accomplished with an iterative procedure—find the time-domain waveform that has the same spectrogram as the original spectrogram. There is a bit more involved in getting the phase from each cochlear channel to line up, but the result is a sound that sounds pretty close to the original. The inversion problem was solved. This also demonstrates that the correlogram representation is a complete model—little information is lost since the sound from the correlogram is so similar to the original.

3 GROUPING CUES

How does the perceptual system understand an auditory scene? Researchers often think about a number of cues that allow the brain to group pieces of sound that are related (Bregman 1990). But this outlook is inherently a bottom-up approach—only cues in the signal are used to perform grouping. There are other cues that come from a person's experiences, expectations, and general knowledge about sound. This chapter argues that both sets of cues are important for scene analysis. First it is useful to review a range of low-level and high-level grouping cues.

3.1 Low-Level Grouping Cues

Many cues are used by the auditory system to understand an auditory scene. The most important grouping principle is *common fate*. Portions of the sound landscape that share a common fate—whether they start in synchrony or move in a parallel fashion—probably originate from the same (physical) object. Thus the auditory system is well served by grouping sounds that have a common fate.

There are many cues that suggest a common fate. The most important ones are common onsets and common harmonicity. Common onsets are important because many sounds start from a common event and all the spectral components are stimulated at the same time. Common harmonicity is important because periodic sounds—sounds that repeat at a rate between 60 and 5000Hz—have many spectral harmonics that vary in frequency and amplitude in a synchronous fashion. These and other similar cues are well described in Bregman's book.

A different style of low-level cues have been exploited for sound separation by the machine-learning community (Lee *et al.,* 1997). Known as blind-source separation (BSS), these systems generally assume there are N distinct sources that are linearly mixed and received by N microphones. Portions of a signal from the same source are highly correlated and thus should be grouped. BSS relies on the statistical independence of different sources. It forces signals apart that do not share a common fate.

A further refinement is possible in the case of one-microphone source separation (Roweis 2003). The original BSS problem remains—find statistically independent sources that sum to the received signals. One-microphone source separation assumes that for any one position in the spectral-temporal plane, only one speaker at a time is present. The problems becomes a matter of allocating the portions of the received signal so that each source is not only independent, but fits a model of speech. One-microphone source separation uses a model of the speaker, encoded as a vector-quantization model in Roweis' work, to guide the separation.

3.2   High-Level Grouping Cues

Many types of auditory scene analysis can *not* be done using simple low-level perceptual cues. Listeners bring a large body of experiences, expectations, and auditory biases to their auditory scene analysis. If somebody says "firetr..." then the only real question is whether I'm hearing the singular or the plural of firetruck, and even that information can be inferred from the verb that follows. Yet, we perceive we've heard the entire word if there is sufficient evidence (more about this in Section 4.3).

Consider an auditory example I presented at a workshop in Montreal. A long sentence was played to the audience. During the middle of the word "legislature," a section of the speech was removed and replaced by a cough. Most people could not recall when the cough occurred. The cough and the entire speech signal were perceived by listeners that understood the English language as independent auditory objects. But one non-native speaker of English did not know the word "legislature" and thus heard the word "legi-coughture." His limited ability to understand English gave his auditory system little reason to predict that the word "legislature" was going to come next. This is an example of phonetic restoration (Warren 1970).

A paper titled "A critique of pure audition" (Slaney and Lyon, 1990) talks about a number of other examples where high-level cues can drive auditory scene analysis. These clues include:

Grouping—Think about a collection of whistles. Would these isolated tones ever be heard as speech? In sine-wave speech three time-varying tone are heard as speech (Remez *et al.*, 1984)

Grouping—A click in an African click language is heard as speech during a spoken utterance, but listeners unfamiliar with that language hear the clicks as instrumental percussion during a song.

Vision affects audition—In the McGurk effect, a subject's visual perception of the speaker's lips affects their auditory perception. (See Figure 13.3.)

Audition affects vision—In a simple apparent-motion demonstration, audio clues can cause motion perception in a simple visual display.

Categorical perception—A particular speech waveform is heard as two different vowels depending on the acoustic environment of the preceding sentence (Lagefoged 1989)

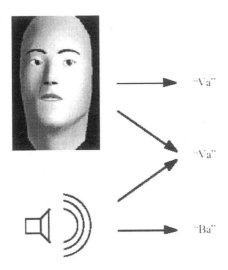

Figure 13.3. The McGurk effect. The same auditory stimulus is heard differently when accompanied by video.

In each of these cases, a listener's experience or a completely different input modality affect the sounds we hear. This high-level knowledge is clearly guiding the scene analysis decisions.

3.3   Which is it? Top-down versus bottom-up

Evidence of the tangled web of perception was described in a paper titled "A Critique of Pure Audition" (Slaney 1996). This paper suggests a number of other effects which call into question a purely bottom-up approach. In the most common cases, simple cues such as common harmonicity causes grouping and we understand the speech. Yet in other cases, speech experience rules the day (i.e. phonetic restoration).

These convoluted connections suggest that sound analysis does not proceed in a purely bottom up fashion. More importantly, how is each sound object represented? With high-level expectations and cross-modality input, it seems difficult to believe that each sound object is represented by a neural spike train corresponding to a real (or hallucinated) auditory waveform.

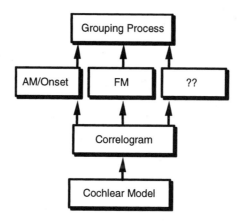

Figure 13.4. The auditory bottom-up grouping process.

4   SOUND SEPARATION MODELS

Researchers use models to concisely describe the behavior of a system. In this section I would like to summarize models that have taken a bottom-up or a top-down view of the world.

Most of the sound separation models to date have evaluated their results based on either the quality of the reconstructions, or the performance of a speech-recognition system on the separate stream outputs. Both approaches largely answer an engineering question: Can we produce a useful auditory scene analysis and improve speech recognition.

The double-vowel perception experiments (e.g. Assman and Summerfield, 1990) are a good example of models that attempt to match the human performance data.

4.1   Bottom-Up Models

Much of the original work on scene analysis used a bottom-up approach that was well articulated by David Marr (1982). In Marr's approach simple elements of a scene (either auditory or visual) are grouped first into simple (2D) cartoons, then more sophisticated processing is applied in steps to create a complete understanding of the object (See Figure 13.4). In this model, the brain performs sound separation and object formation, based on all available clues, before performing sound identification.

The first of the bottom-up models for auditory scene analysis was created by Mitch Weintraub (1986), then a Ph.D. student at Stanford University. In

Weintraub's work, a correlogram is used to analyze the pitches of two speakers (one male and the other female). The pitches were tracked, thus finding the dominant periodicity for each speaker. Then those (spectral) channels from the original cochleagram with the correct periodicity are inverted to recover an estimate of the original speech signal. His goal was to improve speech digit recognition.

In the 1990's a number of researchers built more sophisticated system to look for more cues and decipher more complicated auditory streams. Cooke and Ellis (2001) wrote an excellent summary of the progress to date using correlograms and related approaches to separate sounds. The essential goal is to identify energy in the signal which shares a common fate, group this energy together, and then resynthesize. Common fate identifies sounds which probably came from the same source, often because the energy in different portions of the signal shares a common pitch, or a common onset.

4.2   Top-Down Models

The bottom-up models use information from the sound to group components and understand an auditory scene. Except for information such as the importance of pitch or onsets, there is little high-level knowledge to guide the scene-analysis process. On the other hand, there is much that language and our expectations tell us about a sound.

Perhaps the best example of a top-down auditory-understanding system is a hidden-Markov model (HMM) based speech-recognition system. In an HMM speech-recognition system, a probabilistic framework is built that calculates the likelihood of any given sequence of words given the acoustic observations. This likelihood calculation is based on low-level acoustic features, often based on an acoustic model known as MFCC (Quatieri 2002), but most of the power in the approach is provided by the language constraints.

The complexity of a speech-recognition system's language model is often described by its perplexity, or the average number of words that can follow any other word. Smaller perplexity means that fewer words can follow, the language is more constrained, and the recognizer's job is easier. In a radiological task, the words are specialized and the perplexity is 20, while in general english the perplexity is 247 (Cole *et al.*, 1996). One of the first commercially successful applications of automatic-speech recognition was for medical transcription, where a relatively small amount of high-level knowledge could be encoded as a low-perplexity grammar.

The complexity of the language model directly affects the performance of a speech-recognition system. In the simplest example, once the sounds for "firetr" are confidently heard, then the speech recognition system is likely to recognize the utterance as "firetruck" regardless of what sounds are heard next.

An even more constrained example is provided by the score-following or music-recognition systems (Pardo and Birmingham, 2002). In this case, the system knows what notes are coming and only needs to figure out when they are played. The complexity of a musical signals means that this task can only be accomplished with a very narrow, high-level constraint.

4.3   Mixtures

In practice neither model, top-down or bottom-up, can explain the ability of human listeners to analyze an auditory scene. Clearly low-level cues prime the sound-analysis process—we generally do not hallucinate our entire world. These low-level cues are important, but do not explain simple auditory effects such as phonetic restoration.

Our brains have an amazing ability to hear what might or might not be present in the sound. Consider a slowly rising tone that is interrupted by a short burst of loud noise. As long as the noise burst is loud enough, we perceive a continuous tone that extends through the noise burst, whether the tone is actually present or not. Our brains hear a continuous tone as long as there is evidence (i.e. enough auditory nerve spikes at the correct cochlear channels) that is consistent with the original hypothesis (the tone is present).

This simple demonstration calls into question the location of the hallucinated tone percept. A purely low-level bottom-up model suggests that some portion of the brain has separated out a set of neural spikes that correspond to the phantom tone. The auditory system sees the same set of spikes, with or without the noise burst, and perceives a continuous tone. Instead it seems more likely that somewhere a set of neurons is saying "I hear a tone" and these neurons continue to fire, even when the evidence is weak or non-existent. In other words, object formation is guiding the object segmentation process.

At this point it should be clear that a wealth of information flows up into the brain from the periphery, and a large body of the listener's experience and expectations affect how we understand sound. In the middle information and expectations collide in a system that few have tackled. Grossberg's work (2003) is a notable exception.

The more interesting question, at least for somebody that spent a lot of time developing auditory-inversion ideas, is whether the brain ever assembles a high-fidelity neural coding that represents the pure auditory object. The phonetic restoration illusion only works when the noise signal is loud enough so that the missing speech sound *could* be masked by noise. In other words, the brain is willing to believe the entire word is present as long as it does not violate the perceptual evidence. This is clearly expectation driven, top-down processing. But does the brain represent this missing information as a complete representation of the auditory signal, or use a high-level token to represent the final conclusion (I heard the word "legislature")?

5   CONCLUSIONS

The purely bottom-up approach to auditory perception is clearly inconsistent with the wealth of evidence suggesting that the neural topology involved in sound understanding is more convoluted. One can build a system that separates sounds based on their cochleagram or correlogram representations, but this appears inconsistent with the functional connections. Instead, our brains seem to abstract sounds, and solve the auditory scene analysis problem using high-level representations of each sound object.

There has been work that addresses some of these problems, but it is solving an engineering problem (how do we separate sounds) instead of building a model of human perception. One such solution is proposed by Barker and his colleagues (2001) and combines a low-level perceptual model with a top-down statistical language model. This is a promising direction for solving the engineering problem (how do we improve speech recognition in the face of noise) but nobody has evaluated the suitability of modeling human-language perception with a hidden-Markov model.

A bigger problem is understanding at which stage acoustic restoration is performed. It seems unlikely that the brain reconstructs the full acoustic waveform before performing sound recognition. Instead it seems more likely that the sound understanding and sound separation occur in concert and the brain only understands the concepts. Later, upon introspection the full word can be imagined.

Much remains to be done to understand how humans perform sound separation, and to understand where CASA researchers should go. But clearly systems that combine low-level and high-level cues are important.

6   ACKNOWLEDGEMENTS

Many people helped shape the ideas in this chapter. I am especially grateful for the many wonderful discussions and debates we have had at the Stanford CCRMA Hearing Seminar over the past 15 years. I appreciate the time and energy that Richard F. Lyon and John Lazzaro have spent guiding my thinking in this area. Dan Ellis, in particular, had many useful comments about this chapter.

References

Assman, P.F. and Summerfield, Q., 1990, Modelling the perception of concurrent vowels: Vowels with different fundamental frequencies, *J. Acoust. Soc. Am.* **88**, pp. 680-697.

Barker, J., Cooke M., and Ellis, D.P.W., 2001, Integrating bottom-up and top-down constraints to achieve robust ASR: The multisource decoder. *Presented at the CRAC workshop, Aalborg, Denmark.*

Bregman, A.S., 1990, *Auditory Scene Analysis*, MIT Press, Cambridge, MA.

Cole, R.A., Mariani, J., Uszkoreit, H., Zaenen, A., Zue, V. (eds.), 1996, Survey of the State of the Art in Human Language Technology, *http://cslu.cse.ogi.edu/HLTsurvey/HLTsurvey.html*.

Cooke, M. and Ellis, D.P.W., 2001, The auditory organization of speech and other sources in listeners and computational models, *Speech Comm.*, **vol. 35, no. 3-4**, pp. 141-177.

Grossberg, S., Govindarajan, K.K., Wyse, L.L., and Cohen, M.A., 2003, ARTSTREAM: A neural network model of auditory scene analysis and source segregation. *Neural Networks.*

Ladefoged, P., 1989, A note on 'Information conveyed by vowels,' *J. Acoust. Soc. Am,* **85**, pp. 2223-2224.

Lee, T.-W., Bell, A., Lambert, R.H., 1997, Blind separation of delayed and convolved sources. In: *Advances in Neural Information Processing Systems,* **vol. 9**. Cambridge, MA, pp. 758-764.

Licklider, J.C.R., 1951, A duplex theory of pitch perception, *Experientia* 7, pp. 128-134.

Marr, D., 1982, *Vision*, W. H. Freeman and Co.

Meddis, R. and Hewitt, M.J., 1991, Virtual pitch and phase sensitivity of a computer model of the auditory periphery. I: Pitch identification, *J. Acoust. Soc. Am.*, **vol. 89, no. 6**, pp. 2866-2882.

Pardo, B. and Birmingham, W., 2002, Improved Score Following for Acoustic Performances *International Computer Music Conference 2002*, Gothenburg, Sweden.

Remez, R.E., Rubin, P.E., Pisoni, D.B. and Carrell, T.D.,1981, Speech perception without traditional speech cues, *Science*, **212**, pp. 947-950.

Quatieri, T.F., 2002, *Discrete-Time Speech Signal Processing: Principles and Practice*. Prentice-Hall.

Roweis, S.T., 2003, Factorial Models and Refiltering for Speech Separation and Denoising, *Proceedings of Eurospeech03* (Geneva, Switzerland), pp. 1009-1012.

Slaney, M. and Lyon, R.F., 1990, A perceptual pitch detector. *Proceedings of the International Conference on Acoustics, Speech and Signal Processing.*

Slaney, M., 1996, Pattern Playback in the '90s, *Advances in Neural Information Processing Systems 7*, Gerald Tesauro, David Touretzky, and Todd Leen (eds.), MIT Press, Cambridge, MA.

Slaney, M., 1998, A critique of pure audition, *Computational Auditory Scene Analysis*, edited by David Rosenthal and Hiroshi G. Okuno, Erlbaum.

Warren, R.M., 1970, Perception restoration of missing speech sounds. *Science*, **167**, pp. 393-395.

Weintraub, M., 1986, A computational model for separating two simultaneous talkers. *Proc. of ICASSP '86.*, **Vol.11**, pp. 81-84.

Chapter 14

Techniques for Robust Speech Recognition in Noisy and Reverberant Conditions

Guy J. Brown
Department of Computer Science, University of Sheffield, United Kingdom
g.brown@dcs.shef.ac.uk

Kalle J. Palomäki
Laboratory of Acoustics and Audio Signal Processing,
Helsinki University of Technology, Finland
kalle.palomaki@hut.fi

1   INTRODUCTION

Although much research effort has been expended on the development of automatic speech recognition (ASR) systems, their performance still remains far from that of human listeners. In particular, human speech perception is robust when speech is corrupted by noise or by other environmental interference, such as reverberation (for example, see Assmann and Summerfield, 2003). In contrast, ASR performance falls dramatically in such conditions (Lippmann, 1997). As several researchers have observed (e.g., Hermansky, 1998), the current limitations of ASR systems might reflect our limited understanding of human speech perception, and especially our inadequate technological replication of the underlying processes.

The robustness of human speech perception can be attributed to two main factors. First, listeners are able to segregate complex acoustic mixtures in order to extract a description of a target sound source (such as the voice of a speaker). Bregman (1990) describes this process as 'auditory scene analysis'. Secondly, human speech perception is robust even when speech is partly masked by noise, or when parts of the acoustic spectrum are removed altogether. Cooke *et al.* (2001) have interpreted this ability in terms of a 'missing data' model of speech recognition, and have adapted a hidden Markov model (HMM) classifier to deal with missing or unreliable features. In their system, a time-frequency 'mask' is employed to indicate whether acoustic features are reliable or corrupted; according to this division, the features are treated differently by the recogniser. Typically, the missing data mask is derived from auditory-motivated processing, such as pitch analysis (e.g., Barker *et al.*, 2001; Brown *et al.*, 2001). Alternatively, the mask can be set according to local estimates of the signal-to-noise ratio (SNR) (Barker *et al.*, 2001; Cooke *et al.*, 2001).

This article reviews two studies that estimate a missing data mask for ASR under two acoustic conditions; recognition of speech in the presence of heavy reverberation, and recognition of speech in the presence of mild reverberation and another competing voice. The reader is referred to the two papers concerned (Palomäki *et al.*, 2004a; Palomäki *et al.*, 2004b) for full details.

2   THE MISSING DATA APPROACH TO AUTO-MATIC SPEECH RECOGNITION

The speech recogniser used in both studies employs the missing data technique (Cooke *et al.*, 2001), in which a hidden Markov model (HMM) system is adapted to deal with missing or unreliable data. The classification problem in speech recognition involves the assignment of an acoustic vector Y to a class W, such that the posterior probability $P(W \mid Y)$ is maximised. However, when a noise intrusion is present or when the speech is corrupted by environmental conditions such as reverberation, some components of Y are likely to be unreliable or missing. In these cases, the acoustic model $P(Y \mid W)$ cannot be computed as usual. The 'missing data' technique addresses this problem by partitioning Y into reliable and unreliable components, Y_r and Y_u.

In the simplest approach, the unreliable components are simply ignored, so that classification is based on the marginal distribution $P(Y_r \mid W)$. However, when Y is an acoustic vector it is usually known that the uncertain components have bounded values, and this information can be exploited during classification using the so-called 'bounded marginalisation' method (Cooke *et al.*, 2001). Here, we use bounded marginalisation in which Y is an estimate of auditory nerve firing rate, so the lower bound for Y_u is zero and the upper bound is the observed firing rate. In practice, the recogniser is provided with a binary *mask*, which represents the time-frequency distribution of reliable and unreliable components.

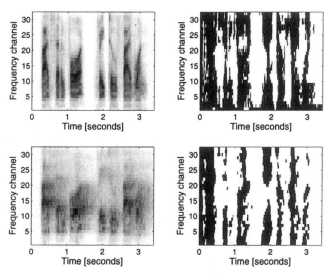

Figure 14.1. A. Rate map computed for the male utterance 'five seven four three two five one' recorded in anechoic conditions. Energy is mapped to gray level: the most energetic regions are darkest. B. The same utterance shown in panel A, but with added reverberation (T60=1.2 sec.). C. The *a priori* mask computed for the rate map in panel B. Black areas in the mask correspond to reliable speech regions, white areas correspond to reverberation contaminated regions. D. Mask computed for the rate map in panel B by a reverberation masking algorithm.

3   MASK ESTIMATION FOR REVERBERANT CONDITIONS

In this approach, modulation filtering is used to identify speech features that are least contaminated by reverberation, and hence to derive a 'reverberation mask' for missing data ASR using spectral features.

The envelope of each channel in an auditory model is processed with a finite impulse response (FIR) filter consisting of a linear phase lowpass component and a differentiator. The filter has a pass band (indexed by 3 dB points) between 1.5 Hz and 8.2 Hz. The aim of this filtering scheme is to detect regions of reverberated speech in which direct sound and early reflections dominate, and to mask the areas that contain strong late reverberation. The role of the lowpass component is to detect and smooth modulations in the speech range, whereas the differentiator emphasizes abrupt onsets, which are likely to correspond to direct sound and early reflections.

Subsequently, a threshold is applied to the modulation-filtered rate map in order to produce a binary mask for the missing data speech recogniser. The value of the threshold should depend on the degree to which the speech is reverberated. In our previous work the threshold was hand-tuned to each reverberation condition (Palomäki *et al.*, 2002), but more recently we have developed a technique for estimating its value directly from an utterance (Palomäki *et al.*, 2004b). Specifically, the threshold is set according to a simple 'blurredness' metric, which exploits the fact that reverberation tends to smooth the rate map by filling the gaps between speech activity with energy originating from reflections.

The reverberation mask estimation process is illustrated in Figure 14.1. The left side of this figure shows an auditory spectrogram ('rate map') for speech in anechoic (A) and reverberant (B) conditions. Missing data masks are shown on the right side of the figure. By comparing the two rate maps pixel-by-pixel, an *a priori* mask (C) can be obtained, in which black pixels correspond to reliable regions (i.e., time-frequency regions whose values did not appreciably change between anechoic and reverberant conditions). Hence, the *a priori* mask may be regarded as a 'ground truth' which indicates the parts of the rate map that have been least affected by reverberation. Panel D shows a mask estimated by the proposed algorithm, which agrees quite closely with the *a priori* mask (C).

The proposed technique has been evaluated on a connected digit recognition task using the Aurora 2 corpus, and compared against a baseline system described by Kingsbury (1998). The latter is a hybrid hidden Markov model-multilayer perceptron (HMM-MLP) architecture, which uses features based on the modulation filtered spectrogram and perceptual linear prediction.

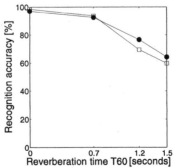

Figure 14.2. Speech recognition accuracy (%) for the proposed reverberation masking system (filled circles) and Kingsbury's hybrid HMM-MLP recogniser (open squares). Test data was 1000 utterances drawn from the test set of the Aurora corpus (male and female speakers). Acoustic models were trained on clean (unreverberated) signals from the Aurora training set. Twelve word-level models were used (1-9, 'zero', 'oh' and silence). Test utterances were convolved with (real) room impulse responses with T60 reverberation times of 0.7 sec., 1.2 sec. and 1.5 sec. In the 0 sec. condition utterances were dry (unreverberated). Data from Palomäki *et al.* (2004b).

Results suggest that the performance of our system is comparable with that of Kingsbury, and even a little better for long reverberation times (Figure 14.2.) Also, we believe that an advantage of our approach is that assumptions about the noise conditions are restricted to the mask estimation rule, which can be adjusted dynamically. On the other hand, Kingsbury's system (and others that use noise-robust acoustic features) are intended to operate across a variety of noise conditions, but are not able to respond dynamically to changes in the acoustic environment.

4 MASK ESTIMATION IN THE PRESENCE OF AN INTERFERING TALKER

For ASR in the presence of an interfering talker at a different spatial location, we employ a system which is divided into monaural and binaural pathways (Figure 14.3.) The monaural pathway is responsible for peripheral auditory processing, and produces feature vectors for the speech recogniser. The binaural pathway is responsible for sound localisation and separation according to common azimuth. Acoustic input to the model is obtained by spatialising speech and noise signals using a model of small room acoustics and head-related impulse responses (HRIRs) to model the filtering effects of the pinnae, head and body.

In the binaural pathway, the temporal fine structure in each frequency channel of an auditory model is processed by a simple model of the precedence effect (see Litovsky *et al.* (1999) for a review). The envelope of the channel is computed, and this is smoothed and delayed to produce an inhibitory signal which is subtracted from the temporal fine structure. The effect is

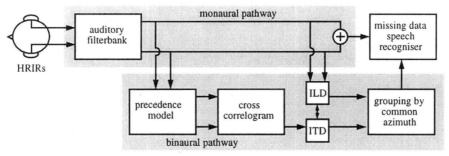

Figure 14.3. Schematic of the binaural auditory model for mask estimation according to spatial location. The monaural pathway generates acoustic features for speech recognition; the binaural pathway estimates a time-frequency mask in which selected regions correspond to acoustic features dominated by the target speaker.

to allow a strong response to direct sound, while the response to late reverberation is inhibited.

Following this, the azimuth of each sound source is identified by a cross-correlation analysis of interaural time difference (ITD). Having identified the location of each sound source, mask estimation is performed by grouping time-frequency regions that share a common azimuth. For frequency channels up to 2800 Hz, this is done based on ITD; each channel of the cross-correlogram is checked to determine whether its response is dominated by the target sound source. For frequency channels above 2800 Hz, the classification is based on interaural level difference (ILD) (see Palomäki *et al.* (2004a) for details). If a time-frequency element is dominated by the target source it is classified as reliable, and the corresponding mask value is set to one; otherwise, the mask is set to zero. Finally, the mask from the binaural pathway and the acoustic features from the monaural pathway are passed to the missing data speech recogniser for decoding.

The proposed binaural separation system was evaluated on a connected digit recognition task, using a subset of 240 male utterances from the TIDigits database (Figure 14.4.) Target speech and another male speaker were arranged at azimuths of (-20, 20), (-10, 10) and (-5, 5) degrees respectively, giving overall spatial separation of 40, 20 and 10 degrees between the two speakers. Target and interfering speech were presented at signal-to-noise ratios (SNRs) of 0, 10 and 20 dB. In all conditions, the proposed binaural system outperformed a baseline HMM recogniser using mel-frequency cepstral coefficients (MFCCs) with deltas, double deltas and cepstral mean normalisation. As anticipated, the performance improvement was most substantial for larger angular separations between the two speakers.

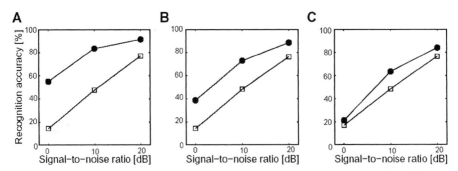

Figure 14.4. Speech recognition in the presence of an interfering talker, which is separated spatially from the target voice by (A) 40 degrees (B) 20 degrees (C) 10 degrees. Recognition accuracy is shown for a set of 240 male utterances from the TIDigits corpus at SNRs of 0, 10 and 20 dB. Filled circles show the performance of the proposed binaural separation system, open squares show the performance of a baseline system which uses MFCCs. Data from Palomäki *et al.* (2004).

5 SUMMARY AND DISCUSSION

Two techniques have been described for estimating a mask for missing data speech recognition, one which is optimised for recognition of speech in the presence of heavy reverberation, and another which is optimised for recognition of speech in the presence of mild reverberation and another speaker. Both systems show performance which exceeds baseline HMM speech recognisers, and the performance improvement is considerable in some conditions.

It should be noted that our binaural separation system may be at odds with some psychophysical findings regarding the role of ITD in concurrent sound separation, since we employ ITD for both simultaneous (across-frequency) and sequential (across-time) grouping. For example, Hukin and Darwin (1995) have shown that listeners only exhibit a weak tendency to segregate a harmonic from a vowel, when that harmonic is given a different ITD to the remaining components of the vowel (see also Culling and Summerfield, 1995). Hence, it appears that across-frequency grouping is primarily mediated by other cues, such as harmonicity. Future work will address this issue by integrating harmonicity and common onset cues into our system; this might give rise to further performance gains, particularly since harmonicity is known to be a relatively robust cue for auditory grouping in the presence of reverberation (Darwin and Hukin, 2000; see also Shamsoddini and Denbigh, 2001). Additionally, we intend to incorporate an equalisation-cancellation mechanism into the binaural system, as an alternative means of estimating time-frequency masks. We expect this to be beneficial in cases where the SNR is very low, or where the spatial separation between target and interferer is small.

Acknowledgments

GJB was supported by EPSRC grant GR/R47400/01. KJP was mainly funded by the EC TMR SPHEAR project and partially supported by the Academy of Finland (project number 1277811) and a Finnish Tekniikan edistämissäätiö grant.

References

Assmann, P. and Summerfield, Q., 2003, The perception of speech under adverse acoustic conditions, in: *Speech processing in the auditory system (Springer handbook of auditory research vol. 18)*, Greenberg, S., Ainsworth, W., eds. Springer-Verlag.

Barker, J., Cooke, M.P., and Green, P.D., 2001, Robust ASR based on clean speech models: An evaluation of missing data techniques for connected digit recognition in noise. *Proc. EUROSPEECH*, **2001**, pp. 213-217.

Bregman, A.S., 1990, *Auditory Scene Analysis*. MIT Press, Cambridge, MA.

Brown, G. J., Barker, J., and Wang, D. L., 2001, A neural oscillator sound separator for missing data speech recognition. *Proc. IJCNN* **2001**, pp. 2907-2912.

Cooke, M.P., Green, P.D., Josifovski, L., and Vizinho, A., 2001, Robust automatic speech recognition with missing and unreliable acoustic data. *Speech Comm.*, **34**, pp. 267-285.

Culling, J.F. and Summerfield, Q., 1995, Perceptual separation of concurrent speech sounds: Absence of across-frequency grouping by common interaural delay. *J. Acoust. Soc. Am.*, **98 (2)**, pp. 785-797.

Darwin, C.J. and Hukin, R.W., 2000, Effects of reverberation on spatial, prosodic and vocal-tract size cues to selective attention. *J. Acoust. Soc. Am.*, **108 (1)**, pp. 335-342.

Hermansky, H., 1998, Should recognisers have ears? *Speech Comm.*, **25**, pp. 3-27.

Hukin, R.W. and Darwin, C.J., 1995, Effects of contralateral presentation and of interaural time differences in segregating a harmonic from a vowel. *J. Acoust. Soc. Am.*, **98 (3)**, pp. 1380-1387.

Kingsbury, B.E.D., 1998, *Perceptually inspired signal-processing strategies for robust speech recognition in reverberant environments*. PhD thesis, Univ. California, Berkeley.

Lippmann, R.P., 1997, Speech recognition by machines and humans. *Speech Comm.*, **22**, pp. 1-15.

Litovsky, R.Y., Colburn, S.H., Yost, W.A., and Guzman, S.J., 1999, The precedence effect. *J. Acoust. Soc. Am.*, **106 (4)**, pp. 1633-1654.

Palomäki, K.J., Brown, G.J., and Barker, J., 2002, Missing data speech recognition in reverberant conditions. *Proc. ICASSP, Orlando, 13th-17th May*, pp. 65-68.

Palomäki, K.J., Brown, G.J., and Wang, D.L., 2004a, A binaural processor for missing data speech recognition in the presence of noise and small-room reverberation. *Speech Comm.*, in press.

Palomäki, K.J., Brown, G.J., and Barker, J., 2004b, Techniques for handling convolutional distortion with 'missing data' automatic speech recognition. *Speech Comm.*, in press.

Shamsoddini, A. and Denbigh, P.N., 2001, A sound segregation algorithm for reverberant conditions. *Speech Comm.*, **33**, pp. 179-196.

Chapter 15

Source Separation, Localization, and Comprehension in Humans, Machines, and Human-machine Systems

Nat Durlach

Hearing Research Center, Boston University, Boston, MA and
Sensory Communication Group, Research Laboratory of Electronics, Massachusetts Institute of Technology, Cambridge, MA
durlach@mit.edu

Collaborators:
Steve Colburn, Gerald Kidd, Chris Mason, Barbara Shinn-Cunningham, Tanya Arborgast, Sasha Devore, and Erick Gallun
Hearing Research Center, Boston University, Boston, MA
Pat Zurek and Jay Desloge
Sensimetrics Corporation, Somerville, MA

1   INTRODUCTION

At the most general level, the task being addressed in this book is that of processing the sound field in realistic acoustic environments (containing both multiple sources and reverberation) in a manner that facilitates goal-oriented behavior in these environments. In terms of classical terminology (most frequently used when human processing rather than machine processing is considered), the task is concerned with "Auditory Scene Analysis" and with factors underlying the "Cocktail Party Effect". Obviously included as sub elements of the specified task (beyond mere detection) are the tasks of (a) separating and localizing the acoustic sources, (b) distinguishing between characteristics of the received signals that are associated with the transmitted signals and those that are associated with the signal transformations (the filtering) imposed by the acoustic environment in which the sources and sensors are located, and (c) comprehending the sources (e.g., understanding the speech and determining the identity of all the talkers in the event that the sources consist of a number of people talking simultaneously).

Although some of the relevant research is sufficiently general to apply to sources other than speech, the focus in this book is on speech communication. However, the types of systems to be considered with respect to the problem of speech reception in complex environments are very broad [including humans, machines (robots), and combinations of humans and machines (human-machine systems)]. Such broad coverage is necessitated both by the multiple goals of the book (understanding humans and designing better machines) and by the belief that the productivity of research in this area can be greatly enhanced by the use of multidisciplinary teams attacking arrays of related problems in a common framework. Overall, it is anticipated that the kinds of research discussed in this book will advance both (1) our understanding of how humans solve the problems in question using their natural biological systems and (2) our ability to design improved machines or human-machine systems for solving these problems. Past research has clearly demonstrated that understanding human processing can benefit design of artificial processing and that knowledge of machine processing techniques can play an important role in understanding human processing. It is also clear that judicious combinations of human and machine processing can lead to systems that are superior to either type by itself, and that design of optimum human-machine systems requires improved knowledge of the advantages and disadvantages of both types of systems.

This chapter is divided into two parts. The first part contains some general comments about the tasks and the systems. The second focuses on two important highly relevant, currently active, research areas in human auditory perception: (A) reverberation and (B) informational masking.

2   GENERAL COMMENTS

Roughly speaking, a given system is said to have separated the sources present in a given acoustic environment if and only if the sounds produced by these sources do not mask each other and there is no confusion about which sound comes from which source (i.e., the system is able to detect all the sounds and to group them into appropriate "auditory objects"). Similarly, localization refers to the ability of the system to correctly identify the location of each source (usually in egocentric coordinates). Finally, comprehension requires that the source message be "understood" (e.g., in the case of speech, that intelligibility is high).

The extent to which and manner in which these tasks can be accomplished obviously depends on the acoustic environment and on the type of system considered (human, machine, or human-machine).

In general, there is a tendency among many investigators to revere human processing and to try and mimic this processing in the design of machines. Although we believe (as indicated in the Introduction above) that machine designers can gain important insights by considering how humans accomplish the various tasks, it is important to keep in mind the many ways in which human processing is poorly matched to some of these tasks. For example, the peripheral human sensing system is obviously not well designed to perform spatial analysis (using only two ears with a limited span, performing frequency analysis prior to spatial analysis, etc.). It would be a simple matter to construct a machine that does a much better job than the human in exploiting spatial separation of the sources to achieve source separation. Similarly, the human system is severely limited with respect to memory (storage) capability (human memory for sounds is strongly limited by trace decay and context-coding noise) and with respect to information input capacity (as exemplified by the extreme difficulty one has comprehending more than two simultaneous speech sources even when direct masking is minimal).

The argument that human auditory processing must be wonderful, like the argument that human speech production must be wonderful, as a consequence of biological evolution, is usually misapplied. Although it may be correct that auditory speech processing and oral speech production constitute superb solutions within the relevant biological constraint space, it is not at all clear that they are such great solutions within the restricted constraint space usually considered by speech and hearing scientists. Thus, as indicated above, although the human sensing array may be near optimal when all the relevant biological constraints are considered (e.g., the constraint on head size imposed by birthing considerations), it clearly is not optimal when these constraints are eliminated. Similar logic applies to the speech-production mechanism. There is no reason to believe that the speech-production system couldn't be improved if one eliminated "irrelevant" evolutionary constraints (e.g., the need to use the oral mechanisms for breathing, eating, and biting one's enemy, as well as for producing speech). In a sense, one can view the notion of using our brains to build machines that are better than the human in this area (as in any other area) as the most relevant evolutionary path. The challenges currently facing machine designers in the effort to achieve good performance in the given task areas are multifold. A list of some of these challenges (in no particular order) is given in the following paragraph.

- Design source-separation processing in such a manner that it does not degrade source comprehension. That this goal is non-trivial is indicated by the extent to which previous work on improving signal-to-interference energy (or power) ratio in speech processing for human listeners using single microphone systems failed to improve (and sometimes degraded) speech intelligibility because of distortions in the target-speech signal caused by the processing.

- Expand the comprehension goal to include identification of speaker and of emotional state as well as high intelligibility.

- Develop processing that works well in realistic acoustic environments containing background interference and multipath propagation (reverberation) as well as unconstrained speech materials (in terms of content, talkers, utterances, etc.).

- Develop systems that not only reduce the degrading effects of multipath propagation on reception of the source message, but clearly separate source characteristics from multipath characteristics of the space and make use of these multipath characteristics to identify the space.

- Optimally integrate various methods of achieving source separation based on spectral-temporal filtering (signal characteristics of sources), spatial filtering (locations of sources), and independent component analysis (statistical relations among sources).

- Fully exploit non-acoustic channels as well as the acoustic channel in source separation and comprehension (e.g., optical lip reading).

- Improve performance in the various tasks through top-down processing and the use of high-level a priori knowledge about speech, talkers, room acoustics, topics under discussion, etc.

- Develop evaluation procedures that adequately measure the performance of a system, provide insight into how the system can be improved, and are applicable to a wide variety of systems.

Although both humans and machines are capable of doing certain components of the specified tasks with at least a modest degree of proficiency, it is obvious that combinations of humans and machines (particularly human-machine systems in which machine processing is followed by human processing) should be able to do much better than either type of system alone. A primary challenge related to the development of human-machine systems is to better delineate the relative advantages and disadvantages of humans and machines for the given tasks (the above comments related to this delineation are merely illustrative).

The main purpose of human-machine systems is twofold: to return human hearing that has been degraded to normal and to create supernormal listening systems (SLS's). The types of degradations that occur can be divided into two classes: "internal" hearing impairments associated with biological degradations (caused by disease, noise exposure, or old age) and "external" hearing impairments associated with the occurrence of interference prior to reception of the acoustic energy at the eardrums (e.g., caused by the wearing of sound-obstructing headgear). Constraints on SLS performance imposed by limitations of the human system focus on human input capacity and human learning rates. Assume, for example, that a machine preprocessor has been constructed that separates and localizes ten simultaneous sources perfectly but is not capable of performing the cognitive functions required to effect comprehension of all the sources. How is it possible to present these ten streams of information to the human system in a satisfactory manner? Even if the notion of "separate ears for separate sources" worked, the human does not have enough ears to go around. At present, the main approaches to this problem, all of which are problematic, are to (1) record and listen to the different sources sequentially, (2) employ team listening (i.e., use one human for each source) or (3) process the multichannel output of the machine in such a way that a single listener is presented with a single source as "foreground" and the other sources as "background" (with an option for the single human listener to select which source is presented as foreground). As one might imagine, efficient use of single-listener real-time systems (such as foreground/background systems) may require significant listener training because of unfamiliar display codes employed or the need to adapt to various sensorimotor transformations.

An additional highly related area in human-machine system research that continues to offer substantial rewards focuses on the time-varying or situation-varying use of humans to help machines or machines to help humans when the other type of system requires help. One such case involves the use of humans to monitor and correct or facilitate machine processing when the machine processing makes errors or encounters problems that are too difficult for the machine to handle alone. In principle, the determination of when the human needs to be called in to help the machine can be made by the machine as well as by the human. In general, applications of this kind can provide efficient and productive (more or less continuous) interpolations between machine processing and human processing.

3   TWO IMPORTANT RELEVANT RESEARCH AREAS IN HUMAN AUDITORY PERCEPTION

3.1   Reverberation

A major thrust in the effort to characterize, understand, and model auditory perception in realistic acoustic environments focuses on the effects of multi-path propagation (reverberation). A crude outline of these effects is shown in Fig. 15.1. As far as physical acoustics is concerned, reverberation tends to increase the overall amplitude of the signals, to modulate the signal in frequency and to smear it in time, and to be directionally diffuse (tending towards isotropicity). In addition, the direct-to-reflected energy ratio (D/R) tends to decrease with distance between source and receiver, and the correlation between the signals received at different locations tends to decrease as the D/R ratio decreases. A pictorial representation of various acoustic effects is shown in Fig. 15.2.

Outline of Effects

- Physical Acoustics (Linear filter that varies with locations and/or orientations of sources and sensors)
    - Amplitude effects (increased energy, direct plus reflected)
    - Temporal effects (smearing, buildup/decay)
    - Frequency effects (alterations of spectra)
    - Direction effects (isotropic)
    - Distance effects (D/R ratio)
    - Decorrelation effects (inter-sensor incoherence)

- Basic Perceptual Effects
    - Positive for absolute detection
    - Mixed for localization
        Hurts direction (but precedence effect)
        Helps distance (D/R ratio)
    - Mixed for monaural masked detection
    - Negative for binaural interaction enhancement
    - Negative for intelligibility (but audibility vs smearing trade)
    - Mixed for subjective effects (binaural dereverberation, live vs dead music, etc)
    X No adequate data on effects of reverberation on source separation?
    X No adequate data on (source, filter) separation?

Figure 15.1. A summary of reverberation effects.

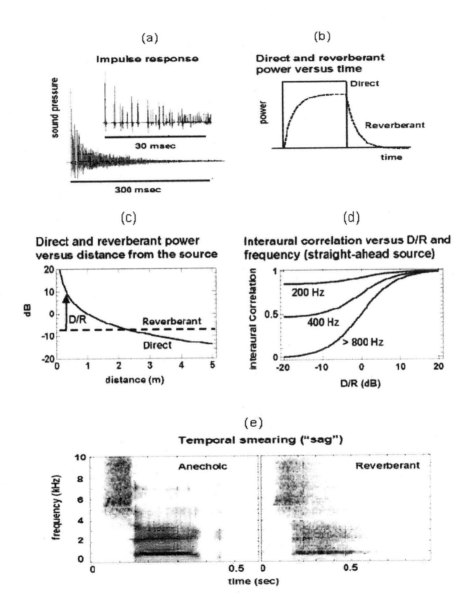

Figure 15.2. Illustrations of reverberation effects: (a) a sample source-to-listener impulse response in a reverberant room (the inset shows a finer time scale); (b) the time-course of direct and reverberant signal power when a sound source is gated on and off; (c) the variation in direct and reverberant power with distance from the source; (d) the interaural correlation as a function of the direct-to-reverberant power ratio; (e) a comparison of spectrograms of the word 'sag' spoken in anechoic and reverberant environments (note the different time scales for the two spectrograms).

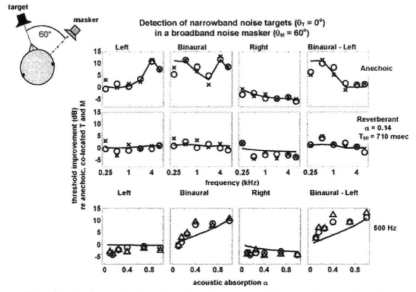

Figure 15.3. Relative level of a broadband masking noise (M) needed to mask a third-octave target signal (T) using stimuli constructed from a computer simulation program and presented through earphones. The simulated room was rectangular with surfaces having a constant absorption coefficient α. The listener was modeled by a rigid sphere with two sensing points simulating the ears. The simulated target source T was directly ahead of the simulated listener with the masker M 60 degrees to the listener's right. Threshold masker levels measured when using the left ear only, both ears (binaural), or the right ear only are plotted in the first three columns; the fourth column shows the difference between binaural and left-ear listening. The top and middle rows show results for α = 1.0 (anechoic) and α = 0.14, respectively, as a function of target center frequency. The bottom row shows the dependence of results on the absorption coefficient α at a fixed target frequency of 500 Hz. All measurements and predictions are normalized to the masker level needed for left-ear listening with co-located target and masker at 0 degrees in an anechoic room. Data points are results from individual listeners; lines are predictions based on statistical room acoustics and the assumption of a constant target-to-masker ratio at threshold for monaural detection; predictions for binaural detection are derived from the EC model. [Reprinted with permission from Zurek *et al.* (2004). "Auditory target detection in reverberation," J. Acoust. Soc. Am. 115, 1609-1620, Copyright 2004, Acoustical Society of America.]

On the perceptual level, these physical characteristics cause reverberation to have a positive effect on detection in quiet, a mixed effect on localization (degrading the perception of source direction but enhancing the perception of source distance), a mixed effect on monaural detection of target sources in backgrounds of interference (the monaural signal-to-interference ratio depending on the details of the configuration), and most often (but not always) a negative effect on binaural unmasking because of interaural decor-

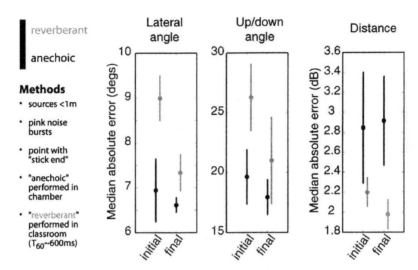

Figure 15.4. Median absolute localization errors (averaged across listeners). Error bars show across-listener standard deviation. Results are compared for a study conducted in a reverberant classroom (see Shinn-Cunningham, Kopco, and Martin, in press, for details about the acoustic environment) and a nearly identical study conducted in an anechoic chamber (Brungart and Durlach., 1999). In both studies, median absolute difference between the source position and the listener response was calculated for the lateral angle, up-down angle, and distance. Median absolute error was computed over the initial 250 trials ("initial") and trials 750-1000 ("final"). Initial errors in lateral and up/down angle were larger in the reverberant room than in anechoic space, but distance errors were significantly smaller in the room than in anechoic space. In anechoic space, errors do not change significantly between the beginning of the experiment and the 1000th trial. In contrast, errors in all three spatial dimensions decrease in the room.

relation. The effect on intelligibility is generally negative because of temporal and spectral alterations; however, speech signals, like other signals, will tend to become more audible. The effects are also mixed subjectively (e.g., consider the effects of different types of reverberation on the enjoyment of music). To the best of our knowledge, there are no adequate data on the effects of reverberation on the task of source separation (as defined in Sec. II above). Also lacking is any serious study of the extent to which and manner in which humans factor the complex spectrum of the signals received at the ears into the complex spectrum of the signal emitted by the source and the complex transfer function associated with the filtering that occurs as a result of the multi-path propagation from the source to the ears.

Data on the effects of reverberation for the detection of a narrowband noise target in a broadband noise masker (from Zurek *et al*, 2004) are shown in Fig. 15.3. The two upper rows of this figure show how the binaural detection advantage that occurs in anechoic space is degraded by reverberation for

Histogram of IPD estimates from running cross-correlation function

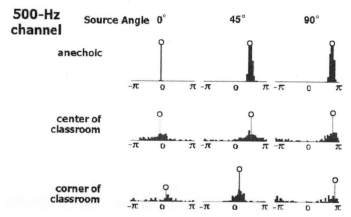

Figure 15.5. Histograms of instantaneous interaural phase difference (IPD) estimates in a wideband noise, extracted from a narrowband 500-Hz short-term cross-correlation model (see Shinn-Cunningham and Kawakyu, 2003). The model input is a broadband noise convolved with head-related impulse responses to simulate the signals reaching the listener's ears in anechoic space or in two locations in a classroom. These acoustic signals are processed through a model of the auditory nerve, then cross-correlated within non-overlapping short (8-ms-long) time windows. The model uses the entire cross-correlation function over interaural time differences from −1 to +1 ms to estimate the IPD in each 8-ms-long time window. The resulting IPD estimates were accrued to produce the histograms in the plots as a function of source angle. The open circle and connecting line in each plot show the best estimate of IPD created by combining the instantaneous estimates over one second. The across-time integration takes into account the IPD of each estimate, as well as its reliability (see Shinn-Cunningham and Kawakyu, 2003). Results show how even modest levels of reverberation increase the variability in instantaneous estimates of source laterality; however, even with this reverberation-caused variability, across-time integration produces reasonably accurate location estimates.

various center frequencies of the narrowband-noise target, whereas the bottom row shows how the degradation depends on the absorption coefficient of the reflecting surfaces. Research is now underway to extend these results (including the theoretical predictions) from the case of detecting a narrowband target signal in environments with both masking noise and reverberation to the case of speech intelligibility in environments with both masking noise and reverberation (Zurek and Freyman, 2003).

The effect of reverberation on errors in the perception of lateral angle, up/down angle, and distance are shown in Fig. 15.4 (Shinn-Cunningham, 2000; Brungart and Durlach, 1999). This figure not only illustrates how reverberation tends to decrease accuracy in the perception of angle (despite the precedence effect) and to increase accuracy in the perception of distance, but also suggests that learning effects may be more pronounced in reverberant conditions than in the anechoic condition (consistent with the notion of "room-learning").

The materials in Figs. 15.5 and 15.6 (which show extensions of the analysis in Shinn-Cunningham and Kawakyu, 2003) illustrate the effects of reverberation on the estimation of interaural phase angle (or time delay), which plays such an important role in both the perception of lateral angle and (according to many theoretical models) binaural unmasking. Fig. 15.5 shows histograms of interaural-phase-difference IPD estimates (for a 500-Hz tone) derived from analysis of running (short-term) cross-correlation functions, together with "final" estimates of these differences based on analysis of all the cross-correlation information. Among other things, these results suggest that the degrading effects of reverberation on the perception of lateral angle can be combated not only by the precedence effect but also by appropriate analysis of the running cross-correlation function (which applies even when the transmitted signal is continuous and does not contain any strong transients). Fig. 15.6 illustrates how the task of detecting a 500-Hz tone at 90° lateral angle in a background of random noise at 0° (using a running cross-correlation model)

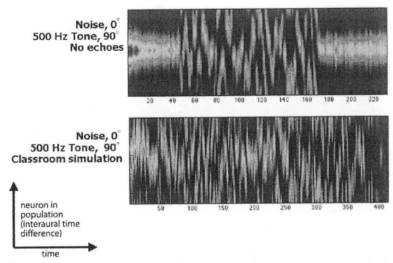

Figure 15.6. Demonstration of the effects of room reverberation on binaural unmasking of a tone in noise. Each panel shows the raw output of the 500-Hz channel of the medial superior olive model described in Shinn-Cunningham and Kawakyu (2003) as a function of time for a wideband noise plus a 500-Hz tone pip. In both panels, the noise has a duration of one second and comes from straight ahead of the listener. The tone has a 100 ms duration and comes 90 degrees to the right of the listener. In anechoic space, the presence of the tone is evident from the across-time variation in the model output, which reflects the interaural decorrelation caused by the tone. However, in a classroom simulation (bottom panel), the interaural decorrelation caused by the reverberation makes detecting the tone decorrelation much more difficult.

is complicated by the presence of reverberation: whereas in the anechoic case (the top strip), the presence of the target tone is obvious because of its clear decorrelation effect, in the reverberant case (the bottom strip), substantial analysis is required to detect the tone because of the decorrelation present even in the noise–alone portion of the stimulus.

The results in Fig. 15.7 illustrate the effect of target-source distance on the target-to-masker ratio (TMR) required for detection in a reverberant environment (relative to the anechoic condition) for the case in which both the target (T) and the masker (M) are at 0° and the case in which the target is moved 90° to the right (Shinn-Cunningham *et al.*, 2002). In the former case, the binaural performance improves with distance because of a decrease in interaural correlation of the target signal with distance (while the masker

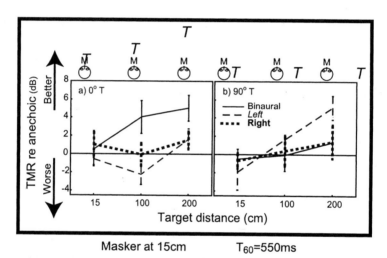

Figure 15.7. Difference in threshold Target-to-Masker Energy Ratio (where threshold TMR is defined as TMR at which target speech sentence is 50% intelligible) between reverberant and anechoic listening conditions (Shinn-Cunningham *et al*, 2002). Plots show the anechoic TMR minus the reverberant TMR. (so that difference is positive when reverberation improves performance. Black lines show results for binaural listening, gray for left ear alone, and black-dashed for right ear alone. Both the masking signal (noise with a spectrum matching the average long-term spectrum of the target utterances) and the target speech were simulated at different locations using head-related transfer functions measured in a classroom. The masking noise was always located 15 cm from the center of the head at azimuth 0 and elevation 0. The target speech location varied; in the left panel, the target was straight ahead, while in the right panel it was at 90 deg azimuth to the right. The abscissa in both panels is the target distance. Results show that reverberation improves binaural conditions (black line, left panel) when the target is more distant than the masker. When the target is to the right, the main effect of reverberation is to improve performance when listening to the left ear stimulus alone, an effect that grows with source distance. This result is consistent with the fact that reverberation effectively boosts the audibility of the left ear signal, which receives relatively little direct sound energy in the anechoic condition.

exhibits high interaural correlation). In the latter case, the left-ear performance improves with distance because of the increased target-energy at the left ear (relative to the anechoic case).

Finally, the results in Fig. 15.8 illustrate the often cited fact that speech reception remains relatively robust for humans under at least moderate reverberant conditions. According to the data shown in this figure (Devore *et al*, 2003; Devore and Shinn-Cunningham, 2003), performance remains over 90 percent correct for the identification of initial and final consonants in nonsense syllables for monaural as well as binaural listening in both a classroom environment and a bathroom environment (chance performance in these tests was 11-percent correct). Apparently, in order for performance to be seriously degraded, both highly reflecting surfaces and a long reverberation time are required. It should also be noted that although consonant identification in quiet is relatively robust for the kinds of reverberation employed in these experiments, when noise as well as reverberation is present, performance can be degraded very substantially, i.e., the effects of reverberation plus noise are worse than the sum of these effects in isolation (an important phenomenon not shown in the accompanying figures).

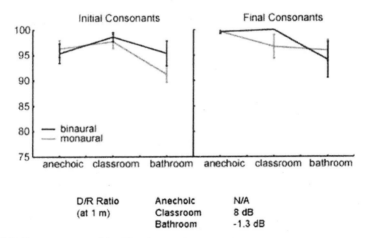

Figure 15.8. Percent correct identification performance for stop consonants in CVC carrier phrases in three simulated environments: anechoic space, a classroom, and a highly reverberant bathroom (Devore *et al*, 2003; Devore and Shinn-Cunningham, 2003). Results are shown for both monaural and binaural listening conditions. The task was to identify the initial (left panel) and final (right panel) consonants in the CVC. Chance performance was 1/8. In all conditions, listeners performed well. Performance was slightly worse in the final consonant conditions in the reverberant environments, where the reverberation due to the preceding part of the CVC utterance helped to mask the final consonant. Binaural listening yielded better performance than monaural performance for initial consonants in the highly reverberant bathroom and for final consonants in the classroom condition.

Generally speaking, reverberation constitutes a major challenge both in our attempt to understand human auditory processing and in our attempt to build machines that function well in realistic acoustic environments. Although significant progress is being made in both of these efforts, much further work is required before success can be claimed for either effort.

3.2   Informational Masking

A second major component of the effort to characterize, understand, and model human auditory perception in realistic acoustic environments that is directly relevant to this chapter focuses on the effects of acoustic interference (masking). Of special interest at the present time are masking phenomena that appear related to central auditory processing. Masking that results from competition between target and masker at the periphery (e.g., in the auditory nerve firing patterns) is now often referred to as "energetic" masking, whereas masking that is not energetic, particularly if it results from stimulus uncertainty, is referred to as "informational" masking. (A recent discussion of some of the relevant conceptual issues in this area, including how one should define "competition at the periphery" and the extent to which one should regard all non energetic masking as informational masking, can be found in Durlach *et al.*, 2003a). While we do not fully understand energetic masking, even after decades of study, the situation is much worse for informational masking: with this latter type of masking, each new finding seems to generate more questions than answers (as illustrated by the puzzles discussed at the end of this chapter).

The experimental results shown in Figs. 15.9 and 15.10 (Durlach *et al.*, 2003b) illustrate (a) the occurrence of informational masking for the case of simultaneous masking of pure-tone signals by multitone complexes with spectra that vary randomly from presentation to presentation (but have no masker energy in the frequency regions near the target so that energetic masking is minimized) and (b) the decrease in informational masking that can be achieved by decreasing the similarity (increasing the dissimilarity) between target and masker. The upper portion of Fig. 15.9 shows schematically how the target signal (indicated by the heavy line) was altered to decrease similarity with the masker [by shortening duration in Exp. 1, by reversing the frequency sweep in Exp. 2, by changing the interaural relations in Exp.3 (spatial), by jittering the target frequency in Exp. 4 (MBS=Multiple-Burst-Same), and by stabilizing the target frequency in Exp. 5 (MBD=Multiple-Burst-Different)]. The lower portion shows, for each of the five experiments, how the amount of informational masking (an amount that can exceed 40 dB in some subjects) was reduced when each of the dissimilarities noted above was intro-

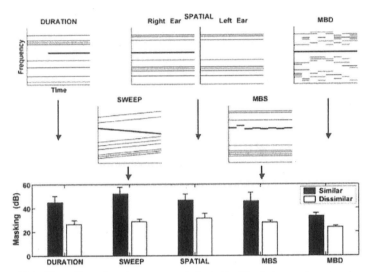

Figure 15.9. The effect of various types of dissimilarities between target and masker on the informational masking of tones masked by multitone maskers with component frequencies selected randomly on each trial (excluding a protected frequency zone around the frequency of the target tone). The upper portion of the figure illustrates the five types of dissimilarity introduced (Duration, Spatial, Sweep, MBS, and MBD) and the lower portion shows the reduction in masking achieved with each type. The error bars show the standard error of the mean (over the five listeners tested). [Reprinted with permission from Durlach *et al*, 2003b, "Informational Masking: Counteracting the effects of stimulus uncertainty by decreasing target-masker similarity", JASA 114, 368-379, Copyright 2003, Acoustical Society of America.]

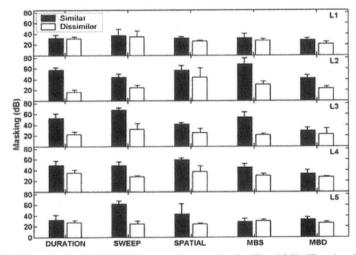

Figure 15.10. Results for individual listeners (see caption for Fig. 15.9). [Reprinted with permission as specified in caption for Fig. 15.9].

duced. Whereas the results shown in Fig. 15.9 represent averages over five listeners, the results shown in Fig. 15.10 are for the individual listeners. These latter results illustrate the large individual differences that occur with informational masking (compare, for example, the results of the MBS experiment for Listener 3 and Listener 5). This raises the obvious question: "Why do subjects differ so much in their susceptibility to informational masking, even when they may differ hardly at all in their susceptibility to energetic masking?" Answering this important question appears to be essential to the development of any comprehensive theory of masking.

The material in Figs. 15.11-17 is taken from a recent research project on informational masking (extending earlier work by Argbogast *et al.*, 2002) in which (a) processed speech signals were used for the target and different, but similarly processed speech signals or random noise were used for the masker;

Constant Conditions:

•CRM sentence corpus: "Ready [call sign] go to [color] [number] now"

•Respond with color (1 of 4) and number (1 of 8) for call sign "Baron"

•15 Frequency bands: filter, extract envelope, modulate sine wave

(Cochlear implant simulation method - convenient - 100% intelligibility)

•Target located at 0°

•Target signal selected randomly on each trial:

8 of 15 bands, 1 of 4 male talkers, 1 of sentences with call sign "Baron"

•Masker used different call signs, colors, numbers

Parametrically Varied Conditions:

•Reverberation: Foam, Bare, Plex

•Masker locations: 0° . 90°

		Masker	Bands	E Mask	I Mask
	Diff-S	CRM Sentence	Different	Low	High
Masker Type:	Diff-N	Noise	Different	Low	Low
	Same-N	Noise	Same	High	Low

Figure 15.11. A brief summary of the conditions tested in the Kidd *et al.* (2004a) study of informational masking. The "Constant Conditions" were common to all experimental conditions while the "Parametrically Varied Conditions" were the main variables under test.

and (b) both spatial separation of sources and room reverberation were manipulated (Kidd *et al.*, 2004a). The processed speech signals were obtained by filtering speech into 15 frequency bands; using the envelope extracted from each band to modulate a carrier with a frequency equal to the center of the band; and then summing over a specified subset of these bands (in all cases, these processed speech signals were highly intelligible). The target signal always consisted of a summation of 8 of the 15 bands with the particular choice of bands (as well as the choice of sentence and talker) selected randomly. The masker then consisted of similarly processed speech using a different (disjoint, non-overlapping) set of bands (Diff-S), random noise using a different set of bands (Diff-N), or random noise using the same set of bands (Same-N). The Diff-N masker was constituted in such a way that its spectrum

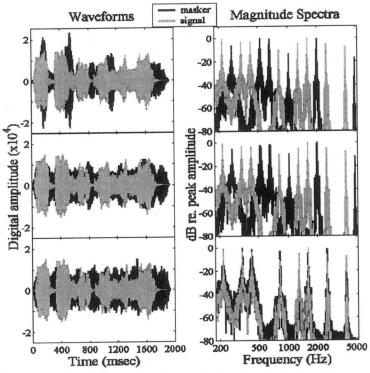

Figure 15.12. Examples of the signals (light gray) and maskers (black) used in the Kidd *et al.* (2004a) study of informational masking. The left column contains plots of the time waveforms while the right column contains plots of the magnitude spectra. The same signal sentence is shown in each row paired with a different-band speech (Diff-S) masker (upper row), different-band noise (Diff-N) masker (middle row) and same-band noise (Same-N) masker (lower row). [Reprinted with permission from Arbogast *et al.*, 2002, "The effect of spatial separation on informational and energetic masking of speech", J. Acoust. Soc. Am.112, 2086-2098, copyright 2002, Acoustical Society of America].

was approximately the same as that of the Diff-S masker. By design, the Same-N masker was predominantly energetic whereas the Diff-S masker was primarily non-energetic. The Diff-N masker was a control for the (very) small amount of energetic masking produced by leakage from the Diff-S bands into the signal bands. The experiments were conducted in a sound field with loudspeakers; the target was always at 0° azimuth with the masker always at either 0° or 90°, and the reverberation characteristics of the room were varied over a wide range of values. The conditions tested were "Bare" (standard IAC booth walls with no acoustic treatments), "Foam" (all surfaces covered with 8" wedges of Silent Source[@] foam), and "Plex" (all surfaces covered with Plexiglas[@] panels). Fig. 15.11 outlines certain conditions of the experiments; Fig. 15.12 illustrates the relation between target and masker for the Diff-S, Diff-N, and Same-N maskers; and Fig. 15.13 provides information on the reverberation present in the conditions Bare, Foam, and Plex.

Some preliminary results from this project are shown in Figs. 15.14 and 15.15. The first of these figures shows the amount of masking evidenced by five listeners for the three different types of maskers, the three different reverberant conditions, and the two signal-masker spatial- separation conditions.

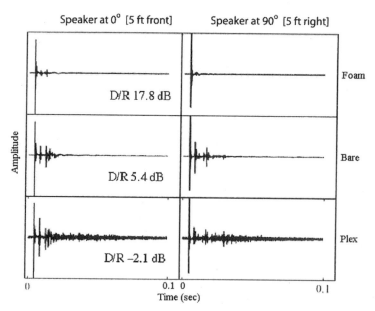

Figure 15.13. Impulse responses for speaker locations at 0 deg (left column) and 90 deg (right column) azimuth for the Foam (upper row), Bare (middle row) and Plex (lower row) room conditions (from Kidd *et al.*, 2004a). [Reprinted with permission from Kidd *et al.*, 2004a, "The role of reverberation in release from masking due to spatial separation of sources for speech identification", Acustica with Acta Acustica, in press.]

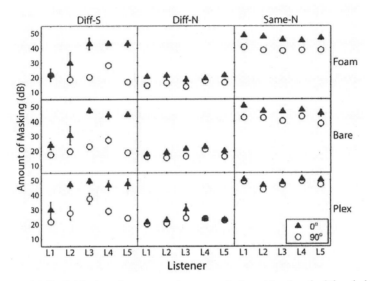

*Figure 15.14.*Individual results expressed as amount of masking (masked threshold minus quiet threshold) for five listeners (L1- L5) for three masker types (columns) and three room conditions (rows). The filled triangles are for 0 deg spatial separation and the open circles are for 90 deg spatial separation (from Kidd *et al.*, 2004a). The error bars show the standard deviation of the results across runs (most of the bars are too small to be seen). [Reprinted with permission from Kidd *et al.*, 2004a, "The role of reverberation in release from masking due to spatial separation of sources for speech identification", Acustica with Acta Acustica, in press.]

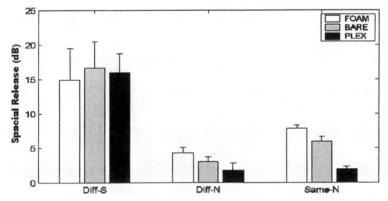

Figure 15.15. Group mean spatial release from masking for the results shown in the previous figure. Spatial release is computed as the threshold at 0 deg separation minus the threshold at 90 deg separation. The results are grouped by masker type (Diff-S, Diff-N and Same-N) for each room condition. The error bars show the standard deviation across listeners. [Reprinted with permission from Kid *et al.*, 2004a, "The role of reverberation in release from masking due to spatial separation of sources for speech identification", Acustica with Acta Acustica, in press.]

The second shows the release from masking in going from the 0° condition to the 90° condition (averaged over listeners). The important results to be noted here about informational masking (the Diff-S conditions) relative to energetic masking (the Same-N condition and the Diff-N condition) are (1) the much greater variability among listeners, (2) the much greater spatial release from masking, and (3) the much smaller degrading effect of reverberation on the spatial release from masking. Presumably, the negative effect of reverberation on the advantages associated with spatial separation in energetic masking is due to the manner in which reverberation corrupts interaural time and amplitude cues. The question then is "Why are the advantages of spatial separation not eliminated by reverberation in informational masking?"

Two further puzzles in the domain of informational masking are represented schematically in Figs. 15.16 and 15.17. In the experiments referred to in Fig. 15.16 (Kidd *et al.*, 2003), the stimuli consist of multiple bursts of sound: the target is a multiple burst of a fixed-frequency tone and the masker is a multiple burst of a multitone complex with frequency components chosen randomly on each trial (outside of a protected region around the target frequency). In the multiple-burst-same (MBS) masker, the component frequencies of the multitone masking complex remain the same for all bursts in the stimulus. In the multiple-burst-different (MBD) masker, these components are jittered in frequency from burst to burst. Generally speaking, detection performance in the MBD case is better than in the MBS case because introducing the jitter into the masker helps the listener separate the target (which is not jittered in frequency) from the masker. As pictured in Fig. 15.16, detection is relatively easy when either the masker is MBD or, if it is MBS, it is moved to the contralateral ear. However when the two conditions are combined (i.e., the MBD masker is in the target ear and the MBS masker is in other ear), detection becomes exceedingly difficult. (Results for the speech analog of this experiment are available in Brungart and Simpson, 2002). It should also be noted that when the masker in either ear is energetic (specifically, Gaussian noise), no additional masking is observed, i.e., the masker in the contralateral ear can be ignored. Thus, an additional issue here concerns the differences between informational and energetic masking with respect to how the effects of the ipsilateral and contralateral masking combine.

In the experiments referred to in Fig. 15.17 (Kidd *et al.*, 2004b), which used the same types of processed speech and noise stimuli as those considered in Figs. 15.11-15, another puzzling phenomenon related to the interaction of maskers and binaural-vs.-monaural stimulation occurs. As indicated in the figure, and as expected, detection is difficult when a Diff-S masker is used to mask a target talker in the same ear, i.e., when informational masking is present. As one might conceivably also expect (although it certainly violates the notion of simple additivity of maskers), detection performance is improved when a Diff-N masker using the same frequency bands as the speech masker is introduced into the same ear. Presumably, this improvement

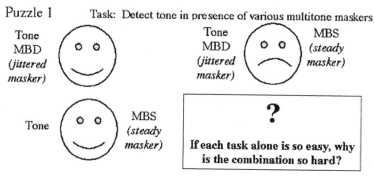

Figure 15.16. A schematic illustrating the conditions tested in the study by Kidd *et al.* (2003). The task is to detect the presence of a sequence of constant-frequency tones in the presence of a sequence of randomized ("jittered") multitone maskers played in the test ear (top left), in the presence of a sequence of constant-frequency tone bursts played opposite the test ear (lower left), and when both conditions are combined (upper right). Lower right poses the paradoxical question motivated by the outcome of this experiment.

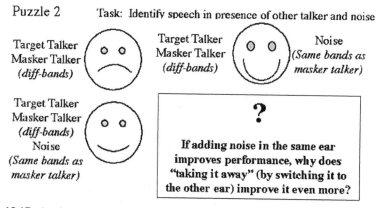

Figure 15.17. A schematic in the same form as the preceding figure showing test conditions in the study by Kidd *et al.* (2004b). The target speech is composed of 8 narrow bands masked by a different talker composed of 6 nonoverlapping bands, a condition that produces substantial informational masking (upper left). When the 6 speech-masker bands are covered up by 6 overlapping narrow bands of noise, the intelligibility of the target improves (lower left) even though more masker energy is present than with the speech masker alone. Paradoxically, shifting the noise masker to the ear opposite the speech target and speech masker improves performance even more (upper right). The question motivated by this unexpected effect is then posed in the panel (lower right).

in detection performance occurs because the masking effect of the noise masker on the speech masker (energetic masking of the speech masker by the noise masker) is greater than its masking effect on the speech target. However, surprisingly, when this same noise masker is switched to the contralateral ear, detection performance improves even more. The puzzling question here then is "Why does the masking effect of the noise masker on the speech masker (relative to its effect on the speech target) have a greater effect when it is presented to the ear opposite to the target than when it is presented to the same ear as the target?"

In general, just as in the case of reverberation, it is painfully obvious that much further work needs to be done to adequately characterize, understand, and model informational masking phenomena.

4   ACKNOWLEDGEMENTS

This chapter is based largely on research performed by the author's collaborators Steve Colburn, Jay Desloge, Gerald Kidd, Chris Mason, Barbara Shinn-Cunningham, Pat Zurek, Tanya Arborgast, Sasha Devore, and Erick Gallun. The author is additionally indebted to Gerald, Chris, Barbara, Pat, and Sasha for their help in preparing the chapter itself. Research support has been provided by the National Institutes of Health, the National Science Foundation, and the Air Force Office of Scientific Research.

References

Arbogast, T.L., Mason, C.R., and Kidd, G. Jr., 2002, The effect of spatial separation on informational and energetic masking of speech, *J. Acoust. Soc. Am.* **112**: 2086-2098.

Brungart, D.S. and Durlach, N.I., 1999, Auditory localization of nearby sources II: Localization of a broadband source in the near field, *J. Acoust. Soc. Am.* **106**: 1956 – 1968.

Brungart, D.S. and Simpson, B.D., 2002, Within-ear and across-ear interference in a cocktail party listening task, *J. Acoust. Soc. Am.* **112**: 2985-2995.

Devore, S. and Shinn-Cunningham, B.G., 2003, Perceptual consequences of including reverberation in spatial auditory displays, *Proc of International Conference on Auditory Displays*: pp. 75-78.

Devore, S., Shinn-Cunningham, B.G., and Durlach, N.I., 2003, Binaural and monaural contributions to spatial release from masking in reverberant and anechoic spaces, *Proc of Mid-Winter Meeting of the Association for Research in Otolaryngology*, Daytona Beach, FL.

Durlach, N.I., Mason, C.R., Kidd, G. Jr., Arbogast, T.L., Colburn, H.S., and Shinn-Cunningham, B.G., 2003a, Note on informational masking, *J Acoust. Soc. Am.* **113**: 2984-2987.

Durlach, N.I., Mason, C.R., Shinn-Cunningham, B.C., Arbogast, T.L., Colburn, H.S., and Kidd, G. Jr., 2003b, Informational masking: Counteracting the effects of stimulus uncertainty by decreasing target-masker similarity, *J. Acoust. Soc. Am.* **114**: 368-379.

Kidd, G. Jr., Mason, C.R., Arbogast, T.L., Brungart, D., and Simpson, B., 2003, Informational masking caused by contralateral stimulation, *J. Acoust. Soc. Am.* **113**: 1594-1603.

Kidd, G. Jr., Mason, C.R., Brughera, A., and Hartmann, W.M., 2004a, The role of reverberation in release from masking due to spatial separation of sources for speech identification, *Acustica* (under review).

Kidd, G. Jr., Mason, C.R., and Gallun, F.J., 2004b, Nonadditivity of energetic and informational masking II: Results from a speech identification task, program of the *2004 Mid-Winter Research Meeting, Association for Research in Otolaryngology*, Daytona Beach, FL.

Shinn-Cunningham, B., 2003, Localization with realistic echoes and reverberation, *Proc of Workshop on Spatial and Binaural Hearing*, Utrecht, the Netherlands.

Shinn-Cunningham, B. and Kawakyu, K., 2003, Neural representation of source direction in reverberant space, *Proc of IEEE Workshop on Applications of Signal Processing to Audio and Acoustics*, New Pfalz, New York, pp. 79-82.

Shinn-Cunningham, B.G., 2000, Learning reverberation: Implications for spatial auditory displays, *Proc of International Conference on Auditory Displays*, Atlanta, GA, pp. 126-134.

Shinn-Cunningham, B.G., Constant, S., and Kopco, N., 2002, Spatial unmasking of speech in simulated anechoic and reverberant rooms, *Proc of 25th Mid-Winter meeting of the Association for Research in Otolaryngology*, St. Petersburg Beach, FL.

Shinn-Cunningham, B.G., Kopco, N, and Martin, T.J., 2004, Acoustic spatial cues contained in reverberant binaural room impulse responses from a classroom, *J. Acoust. Soc. Am.*, under revision.

Zurek, P.M. and Freyman, R.L., 2003, Predicting masked detection and speech recognition in reverberant rooms, *Proc of Mid-Winter Meeting of the Association for Research in Otolaryngology*, Daytona Beach, FL.

Zurek, P.M., Freyman, R.L., and Balakrishnan, U., 2004, Auditory target detection in reverberation, *J. Acost. Soc. Am.* **115**, 1609-1620.

Chapter 16

The Cancellation Principle in Acoustic Scene Analysis

Alain de Cheveigné
Ircam-CNRS, Paris, France
alain.de.cheveigne@ircam.fr

1   INTRODUCTION

The acoustic environment is often cluttered. The ears of an organism sample *mixtures* of acoustical waveforms coming from multiple sources, rather than the source waveforms themselves. Making sense of the environment on this basis is a process known as Auditory Scene Analysis, or ASA (Bregman 1990). If the organism is interested in a particular source (the *target*), the presence of other sources (*jammers*) interferes with target perception. Unfortunately, perceptual models are generally designed to handle a single isolated source, and extending them to work within a complex environment is a challenge. Similar problems arise when designing an artificial device (such as a speech recognizer) to work in an acoustically cluttered environment.

Cues used by humans have been reviewed by Bregman (1990). Generally speaking, they consist in *regularities* of the source and/or the scene. These include spatial location (correlation between ears or sensors), periodicity (correlation across time), common onset (correlation across frequency channels), familiarity (correlation with predetermined templates or patterns), etc. Artificial systems have been built that use similar regularities (Cooke and Ellis 2001). Traditionally, most efforts have concentrated on regularities of the *target* that allow it to be enhanced. This paper describes an approach that instead uses regularities of the *jammers* to suppress them.

Jammer "structure" takes many forms. One or several jammers may be predictable, or periodic, or jammer components may be correlated across several sensors. These basic structures may be extended to include amplitude variation, frequency modulation, moving sources, etc. Each bit of exploitable jammer structure opens a window through which the target can be "glimpsed".

The focus here is mainly on artificial systems (typically automatic speech recognition, ASR), but understanding how the auditory system handles such tasks is also a goal, in itself and as a source of ideas for better algorithms. Conversely, effective algorithms may serve as models to guide our investigation of natural processes.

2   TASK AND CONTEXT

The task is to recognize or recover a target source within a noisy environment. For simplicity, suppose two sources T (the "target") and J (the "jammer") that are observed indirectly from signals X and Y provided by one or two microphones. This structure can be generalized as needed to more sources and/or sensors as needed. Sources and observations are related via a *mixing matrix* that is convolutive: each matrix element is a transfer function (or impulse response) that represents the effects of propagation delay and dispersion from a source to a transducer (Fig. 16.1). Two subtasks are of interest. The first is to derive useful information about the *structure* of the scene and/or the sources: intersensor correlation, source fundamental frequencies (F_O s), etc. The second is to recover a "clean" version of the target.

It is often possible to derive an approximation T' of the target from the observed signals. T' depends on both target and jammer:

$$T' = f(T) + \varepsilon(J) \qquad (1.1)$$

Ideally we'd like $f(\ )$ to be identity and $\varepsilon(\ )$ to be zero (no distortion and no crosstalk, respectively). Arguably of these two ideals the latter is the most useful. Whereas target distortion is typically predictable and can be compensated, crosstalk is usually unpredictable and cannot.

Typical application contexts are ASR, conference systems, hearing aids, musical applications (recording, score following, interactive systems), multimedia indexing, etc.

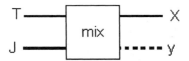

Figure 16.1. Observed signals X (and possibly Y) are related to target T and jammer J via a mixing matrix. The goal is to derive information about the target T.

3 ASSUMPTIONS ON SOURCE AND SCENE STRUCTURE

Cancellation is usually applied in the time domain. At each instant t, an estimate of the jammer waveform is *subtracted* from the compound waveform. Three cases are of interest, that differ according to whether the jammer estimate comes from (1) a predetermined template waveform, (2) previous values of the waveform being processed, or (3) the waveform of another sensor.

The first case (subtraction of a waveform template) is ideal but rare. Examples might be a stationary jammer, or the stereotyped waveform of an instrument note, either known beforehand or estimated from the context. It is ideal because subtraction leaves the target undistorted.

The second case is that of a *periodic* jammer $J_t = J_{t-P}$ where P is the period. Suppose that the observed signal is the sum of the target and the jammer $X_t = T_t + J_t$. By subtracting X_{t-P}, the contribution of J is suppressed:

The result T' depends only on T and not on J. It is spectrally distorted as a result of the processing, but jammer rejection is infinite.

$$T_t = X_t - X_{t-P} = T_t - T_{t-P} \qquad (1.2)$$

The third case is that of multiple sensors in an anechoic environment. Things are a bit simpler if X and Y are rescaled in time and amplitude so that the contribution of J to each is the same: $X_t = \alpha_x T_{t-\tau_x}$, $Y_t = J_t + \alpha_y T_{t-\tau_y}$. This contribution is then suppressed by forming:

$$T_t = X_t - Y_t = \alpha_x T_{t-\tau_x} - \alpha_y T_{t-\tau_y} \qquad (1.3)$$

The result T' depends only on T and not on J. It is spectrally distorted as a result of the processing, but jammer rejection is again infinite.

These basic cases can be extended. For example the periodic jammer model can be extended to a *variable amplitude* periodic jammer ($J_t = \alpha J_{t-P}$). For that, Eq. 1.2 is replaced by $T' = X_t - \alpha X_{t-P}$. A *variable frequency* jammer can be handled by time warping the observed signal before processing, a moving source by a combination of time warping and gain adjustment, etc. These operations may be performed within bands of a filterbank, with coefficients that vary from band to band.

The basic cases can also be combined (e.g. multiple periodic sources picked up by multiple sensors, etc.). Cancellation fails in two cases: (a) the jammer does not fit any structure model, and (b) it does, but the target fits the same model. The rest of this paper discusses how to handle those cases. Before that, we discuss the issue of *estimation* of the source and scene structure parameters.

4   ESTIMATING SOURCE AND SCENE STRUCTURE

Several aspects of the structure of the source or scene are useful for cancellation, supposing that their parameters can be estimated from the available data.

- **Jammer template.** In some cases the jammer waveform can be completely estimated. A simple example is a deterministic stationary jammer such as hum (power frequency harmonics picked up by low-level audio circuits). Granted the mild assumption that the target has intervals of low amplitude, the jammer template can be obtained from a fit to the waveform in those intervals. Granted the further assumption that the jammer is indeed stationary and deterministic, the template is interpolated and subtracted from the entire waveform. The advantage is that the jammer is subtracted, rather than filtered out, and thus there is no spectral distortion of the target. More complex examples are possible but not discussed here.

- **Periodicity.** In other cases the *period* of the jammer, rather than its waveform, can be estimated. Cancellation itself can be used for this purpose. The idea is to search the parameter space of a cancellation filter looking for a *minimum residual output*. For example, to estimate the period of an isolated source the filter defined by Eq. 1.2 is applied and its parameter P is varied until a minimum is found. This principle was applied with success in the YIN method of F_O estimation (de Cheveigné and Kawahara, 2002). The same principle can be extended to multiple sources (de Cheveigné and Kawahara, 1999; de Cheveigné and Baskind, 2003).

- **Intersensor delay/attenuation.** The jammer waveform may lack structure, but it may contribute to several observations with certain delays and attenuation factors. Again, these factors may be estimated by cancellation. The idea is to search the parameter space of a spatial cancellation filter (null beamformer) looking for a minimum of the residual output. For example, to estimate delay and attenuation of a single source supposing nondispersive propagation, the filter defined by $X_t - \alpha X_{t-\tau}$ is applied to sensor signals and its parameters α and τ varied until a minimum is found.

The principle can be extended to *dispersive propagation* and *more than two sources/sensors* by splitting the signals over a filterbank and working within narrowband channels. More on this later. From intersensor parameters one can infer source positions (within surfaces of confusion). However source positions are not of direct use for cancellation unless we wish to include spatial constraints, for example within a multimodal system.

- **Joint estimation.** Periods, intersensor parameters, and templates can be estimated *jointly*. In this case, estimation of each aspect of the structure is aided by other aspects. For example F_O estimation may be aided by spatial structure, and vice-versa.

5 RECOVERING THE TARGET

Supposing the scene fits a structure model, and its parameters are known, a time-domain waveform T' can be obtained according to equations analogous to Eqs. 1.2 and 1.3. This waveform (or its equivalent spectrum) is then fed to a pattern-matching or resynthesis stage, together with structure parameters if needed. As pointed out in Section 2, cancellation allows perfect jammer rejection in ideal conditions. In practice these conditions may arise only within limited time or frequency intervals.

6 LOCAL CANCELLATION AND MISSING DATA

A likely event is that cancellation is possible for a restricted *temporal interval*. For example if the jammer is voiced speech, harmonic cancellation can be applied only during steady-state voiced segments, during which the target may be "glimpsed". Cancellation might also be possible within a restricted *spectral interval*. For example, narrow-band noise may prevent cancellation within some bands. The target is "glimpsed" within the bands that remain. Combining both ideas, one may apply cancellation within a restricted *spectrotemporal region*. Note however that the efficacy of simultaneous bounds in time and frequency is limited by the Gábor relation (Gábor, 1947).

Supposing cancellation is effective only locally, parts of the target will be missing. The parts that remain may nevertheless be sufficient for a task such as pattern-matching (e.g. ASR). *Missing data techniques* have been developed to address this situation (Cooke *et al.*, 1997; Lippmann and Carlson, 1997; Morris *et al.*, 1998). Missing features are either ignored, or (if possible) constrained by bounds derived from the target + jammer mixture. These techniques assume a "mask" to tell them which intervals are missing. In the context of cancellation, the mask is a by-product of the cancellation process.

A second problem is that the target "glimpses" are usually spectrally distorted by the cancellation filters. An option is to compensate by inverse filtering, but a more general solution is to apply similar distortion to the *templates* in the pattern-matching stage. Information needed for that purpose may be available from the cancellation stage. Template (or model) adjustment is not yet common among missing feature techniques (see de Cheveigné, 1993b, for an early attempt).

7   MODELS

Pattern-matching is a special case of *model fitting*. Once a model is fitted (possibly on the basis of incomplete data) it allows *interpolation*. The models embedded in an ASR system (states, covariance matrices, dictionaries, etc.) can be used in this way. Other useful models are articulatory, multimodal, linguistic, etc. *Redundancy* relations between features may allow accurate interpolation when one feature is missing and the other not.

8   POWER AND VARIANCE PARTITION

Obviously one must know which features are reliable and which are not. This section suggests one possible approach to obtain this information from the observed signals. The idea is to partition the *power* within a mixture into parts that reflect various sources. This partition is also useful as a partition of the *power spectrum* (thanks to Parseval's relation). A partition of power can also be interpreted as a partition of *variance* (sum of squares). Variance estimates can then be used to parametrize statistical models from which feature reliability can be inferred.

As an example, consider a quasiperiodic jammer J. It is possible to express it as the sum of two signals J' and J'':

$$J'_t = (J_t - J_{t-P})/2, \qquad J''_t = (J_t + J_{t-P})/2, \qquad (1.4)$$

If J is purely periodic with period P, then $J'=0$ and $J''=J$. J' is nonzero only if J is not perfectly periodic, and in that sense we can call J' the "aperiodic" part of J, and J'' the "periodic" part.

What makes this partition useful is that it is also a partition of power. Defining the local power of a signal X (measured over a window starting at t) as:

$$\|X_t\|^2 = (1/W) \sum_{j=t+1}^{t+W} X_j^2, \qquad (1.5)$$

it is easy to verify that:

$$(\|J_t\| + \|J_{t-P}\|^2)/2 = \|J_t'\|^2 + \|J_t''\|^2 \qquad (1.6)$$

The term on the left is the average of two estimates of the power of the jammer (over slightly different windows), and the right hand terms are powers of aperiodic and periodic parts respectively. Parseval's relation implies a similar partition of *power spectra*. Spectrally, the partition can be represented by the transfer functions $1 - \cos(2\pi fP)/2$ and $1 + \cos(2\pi fP)/2$.

In the context of cancellation, J' represents *crosstalk*. If T' is the cancellation-filtered target, the output of the cancellation stage is $T'+J'$. The quality of the recovered target depends on the relative weights of $\|T\|$ and $\|J\|$. These cannot be observed, but there are several situations where they can be inferred:

15. Jammer properties may be known well enough to put an upper bound on the ratio $\|J'\|/\|J\|$. Using the power of the observed signal $\|X\|$ as a statistically conservative bound on $\|J\|$, we get an upper bound on crosstalk power $\|J'\|$. Thanks to Parseval's relation, this reasoning may be applied to each *frequency*.

16. The target too may be periodic. A full analysis is complicated and will be outlined only briefly. Calling P and Q the periods of jammer and target, the observable signal X can be expressed as the sum of four parts:

$$X_t^1 = (X_t - X_{t-P} - X_{t-Q} + X_{t-Q-P})/4$$

$$X_t^2 = (X_t + X_{t-P} - X_{t-Q} - X_{t-Q-P})/4$$

$$X_t^3 = (X_t - X_{t-P} + X_{t-Q} - X_{t-Q-P})/4$$

$$X_t^4 = (X_t + X_{t-P} + X_{t-Q} + X_{t-Q-P})/4 \qquad (1.7)$$

As above, this defines a partition of signal power. The first quantity X^1 is zero iff target and jammer are perfectly periodic (quantities X^2 and X^3 are zero if target or jammer are periodic, respectively). Under certain assumptions X^1 can be used as an estimate of the power that is "unaccounted" for by a sum-of-periodic-signals model, i.e. crosstalk. Again, this reasoning can be applied to each frequency, based on Parseval's relation.

Similar operations can be performed in the multisensor and hybrid cases. Power is defined as a mean sum of squares, and as such it is equivalent to mean *variance*. Ratios of variance can be interpreted as measuring the uncertainty with which the target is observed within in each frequency band, at each time frame, and thus the power partition offers the opportunity of interpreting observations according to a statistical model.

9   RELATION WITH AUDITORY MODELS

Barlow (1961, 2001) suggested that the role of sensory relays is to recode incoming patterns in a way that minimizes numbers of neural discharges, and thus metabolic cost, on average. Cancellation fits this description. A "neural cancellation filter" (e.g. de Cheveigné 1993a) minimizes its output for a periodic input, and at the same time characterizes the regularity of the input pattern.

Durlach's (1963) equalization-cancellation (EC) model proposed that patterns from one ear are subtracted from those from the other (after delay and amplitude scaling) to suppress correlates of a spatially localized jammer. Culling and Summerfield (1995, Culling *et al.*, 1998) proposed a "modified EC" model in which such cancellation occurs independently within peripheral filter bands. In this model, EC parameters are determined from information within a band, and may differ from band to band. See also Breebart *et al.* (2001) and Akeroyd and Summerfield (2000).

A monaural "harmonic cancellation" model was proposed by de Cheveigné (1993a) and found to account for behavioral data on concurrent vowel identification (de Cheveigné, 1997). In particular it accounted for conditions where one vowel is much weaker than the other, for which other explanations fail. A "cancellation model of pitch perception" was proposed by de Cheveigné (1998). A model that explains pitch shifts of inharmonic partials (Hartmann and Doty, 1996) was proposed by de Cheveigné (1999a). Given the general functional usefulness of cancellation (as argued in this paper) and the fact that some of these models account for effects that no other model accounts for, it is likely that the cancellation strategy is used within the auditory system.

Understanding auditory processes is goal that is worthy in itself. It is also a source for insight into effective processing techniques, and a great opportunity for interaction of mutual benefit between scientific and technological fields. To constrain and develop such useful models, there is strong need for more data on natural systems via behavioral, physiological, and imaging techniques.

10   RELATION WITH OTHER TECHNIQUES

10.1   Time-Frequency Decomposition

Many efforts have been devoted to computational models of ASA (e.g. Cooke, 1991; Brown, 1992; Ellis, 1996). A common approach is to assume a *spectrotemporal decomposition* of each sensor signal over a filterbank, *grouping* together of filter bands that belong to the target, and their *segregation* from bands that belong to other sources. Bands are assigned according to a time-frequency "map" that looks like a checkerboard.

The idea comes from the ASA rules reviewed by Bregman (1990), themselves based on the principle of peripheral frequency analysis that originated with Helmholz (1877). Strict Helmholtzian doctrine would have it that the outputs of the bands collectively form a *spectrum* of slowly-varying values (excitation pattern). Recent thinking, both in auditory models and in CASA systems, allows for each band to carry a *temporal* structure, that may be used to decide how the band is assigned. Early examples are the two-channel system of Lyon (1983), that drew on Jeffress's (1948) localization model to segregate bands according to source bearing. Another is the single-channel system of Weintraub (1985) that drew on Licklider's pitch model to segregate bands according to source periodicity. More recent examples are the CASA systems of Cooke (1991), Brown (1992) or Ellis (1996). Decomposition into time-frequency "pixels" is also used in missing-feature techniques (Cooke *et al.*, 1997; Lippmann and Carlson, 1997; Palomäki *et al.*, 2001), statistical methods for time-frequency pixel assignment (Roweis, 2000, 2003), or multiple F_O estimation (Wu *et al.*, 2003).

There is considerable variety among systems based on time-frequency analysis. Frequency analysis may be performed by a bank of "auditory" filters, by a standard short-term Fourier transform, or by a more exotic time-frequency transform. The output is either a slowly-varying spectrum, or a set of rapidly varying temporal waveforms filtered from the input waveform. At each instant a band is assigned entirely to a source ("black and white" map) or only partially ("gray-scale" map). Common to all systems is that bands are "atomic" in the sense that they are not analyzed further.

The effectiveness of the time-frequency approach is limited by the Gábor relation: $\Delta f \Delta t \leq$ constant. As an example, the response of a 1 ERB wide gammatone filter centered at 1 kHz is only 20 dB down (1% power) at 200 Hz away from the peak. Its impulse response is 20 dB down at 6 ms from the time of peak response. Spectral resolution can be improved only at the expense of temporal resolution, and vice-versa, and so jammer rejection cannot be perfect.

Cancellation is complementary with time-frequency analysis. In ideal conditions it offers perfect jammer rejection, but these ideal conditions may prevail only within a limited time-frequency region. Cancellation cannot be subsumed by time-frequency analysis, but the two approaches are complementary and may usefully be combined.

10.2   Enhancement

Enhancement is the mirror image of cancellation. Rather than using jammer structure, *target* structure (periodicity, spatial position) is used to enhance a structured target relative to an unstructured background. Enhancement schemes are much more common in the literature than cancellation. However the SNR improvement that they provide is generally limited. For delay-and-add beamforming it is 6 dB for two sensors, and greater improvement requires more sensors. For harmonic enhancement it is 6 dB for a simple comb-filter, and greater improvement requires filters with longer impulse responses (de Cheveigné, 1993a, Appendix A). Cancellation is distinct from (and complementary to) enhancement.

10.3   ICA

Independent component analysis and cancellation are related. The objective of ICA is to produce outputs that are statistically independent. This can happen only if each output depends on one source only, a goal that is attained if contributions of all other sources are suppressed. Thus, the objectives of ICA and cancellation are equivalent, even if the means to attain them are different. The links between ICA and cancellation should be examined more deeply, and it may eventually turn out that ICA and cancellation can be subsumed within a common framework.

It is interesting to note the similarity between Culling and Summerfield's mEC model, and recent frequency-domain ICA techniques (e.g. Anemueller, 2001). Both are congruent with the notion of "local" cancellation described in this paper.

11 COMPUTATIONAL CONSIDERATIONS

Estimation of structure parameters using cancellation is expensive, because (except in special cases) the parameter space must be searched exhaustively. *Joint estimation* of several parameters is particularly expensive. Techniques to reduce the cost are described in de Cheveigné (2001).

12 PUTTING IT ALL TOGETHER

Here are a three example scenarii, of varying complexity, of how cancellation might fit together with other techniques to solve a problem of practical interest.

- **ASR system with single channel input.** Cancellation is used for several purposes: (1) for an isolated voice, to provide F_O, F_O-smoothed spectra, and a time-frequency "harmonicity map" as features for ASR, (2) for two concurrent voices, to provide "glimpses" of both voices, together with time-frequency reliability maps for both. These are used by the ASR stages to constrain models of one or more speakers. Spectral distortion caused by cancellation is compensated in the ASR stage by adjusting spectral models.

- **Active multimodal recording system.** A room (conference room or concert hall) is equipped with a distributed network of switchable microphones (or robot controlled microphones) and video cameras. Cancellation is used to analyze the acoustic structure of signals provided by the microphones. The harmonic structure of sources (voices, instruments) is used to facilitate the acoustic analysis. Its result feeds a spatial model that is also informed by video and any other relevant information. The spatial model is used to switch or move microphones, to optimize pickup and segregation of each source of interest, or to produce a visual display of use to the sound engineer. Cancellation analysis reveals that scene structure information is incomplete (for example intersensor correlation may be good only at note onsets, for which the anechoic propagation approximation is good). Incomplete information is interpolated using *missing data techniques* to constrain *models*. Models are then used in the next stage to *interpolate* across missing parts, in the event that the system was incapable of recovering them. Models at all stages, including ASR, can be merged and fit jointly (e.g. Nakamura and Herakleous 2002). On the basis of models, it may be possible to *resynthesize* high quality speech or music sounds (e.g. Kawahara, this volume).

- **Multimedia indexing and search.** A major problem in dealing with massive volumes and fluxes of multimedia data, as they occur today, is indexing and search. The concept of *metadata* has been invented for that purpose. Arguably the most useful kind of metadata are *content-based*, as they are cheap, reliable and ubiquitous (as compared to text and other manually created metadata that are expensive and therefore often absent). Content-based metadata can be used to map out redundancies (e.g. copies of same data) and constrain other forms of metadata. They are essential for efficient search.

  For mixtures of audio sources, it would be desirable that the metadata reflect the sources enough to support searching for individual sources within the metadata that label the mixture. It is usually not possible to split audio data into streams and label each stream. However it is possible to design content descriptors so that they maximize information about component sources. Cancellation is useful for such labeling. As an example, a single channel containing several periodic sources can be processed so as to obtain (a) estimates of the periods, and (b) a periodicity-based decomposition of power and power spectra. It is not necessary that segregation be perfect: anything that allows pruning of the search space is a sufficiently useful goal.

  The power spectrum decomposition is also a decomposition of variance, and thus it fits well with statistical models that support hierarchical search (de Cheveigné, 2002). It also fits well with the scalable metadata concept that has been integrated into the audio part of the MPEG-7 standard (de Cheveigné 1999b; ISO/IEC JTC 1/SC 29, 2001). The additive nature of variance implies that "decomposed" and "standard" descriptions are compatible. Together with the scalability of metadata structures (also based on variance), this ensures interoperability and flexibility of the metadata descriptions.

13   CONCLUSION

Cancellation is a useful "ingredient" to solve the problems of speech separation and acoustic scene analysis. Other essential ingredients are time-frequency analysis, models, and missing-data techniques. The strength of cancellation is that it can provide, in ideal conditions, infinite jammer rejection. Its weakness is that these ideal conditions may occur only locally, in time and/or frequency, hence the need for models and missing-data techniques. This approach should benefit from future progress in signal processing techniques such as beamforming and ICA, and also from being cast into a systematic probabilistic framework. There are arguments to say that neural processing in natural organisms is in part based on cancellation. More basic knowledge is needed about the nature of these mechanisms, their anatomy and physiology, and the behavior that they allow.

14   ACKNOWLEDGMENTS

Thanks to Pierre Divenyi, Dan Ellis and Deliang Wang for organizing this workshop, and for providing the stimulation to work on these ideas. Thanks to NSF for funding to support the workshop

References

Akeroyd, M.A. and Summerfield, A.Q., 2000, A fully-temporal account of the perception of dichotic pitches, *Br. J. Audiol.* **33**(2):106-107.

Anemueller, J., 2001, *Across-frequency Processing in Convolutive Blind Source Separation*, Oldenberg, unpublished doctoral dissertation.

Barlow, H.B., 1961, Possible principles underlying the transformations of sensory messages. in: *Sensory Communication*, W. A. Rosenblith, ed., MIT Press, Cambridge, Mass, pp. 217-234.

Barlow, H.B., 2001, Redundancy reduction revisited, *Network: Comput. Neural Syst.* **12**:241–253.

Breebaart, J., van de Par, S., and Kohlrausch, A., 2001, Binaural processing model based on contralateral inhibition. I. Model structure, *J. Acoust. Soc. Am.* **110**:1074-1088.

Bregman, A.S., 1990, *Auditory Scene Analysis*, MIT Press, Cambridge, Mass.

Brown, G.J., 1992, *Computational Auditory Scene Analysis: a Representational Approach*, Sheffield, Department of Computer Science, unpublished doctoral dissertation.

de Cheveigné, A., 1993a, Separation of concurrent harmonic sounds: Fundamental frequency estimation and a time-domain cancellation model of auditory processing, *J. Acoust. Soc. Am.* **93**:3271-3290.

de Cheveigné, A., 1993b, Time-domain comb filtering for speech separation, *ATR Human Information Processing Laboratories technical report*, TR-H-016.

de Cheveigné, A., 1997, Concurrent vowel identification III: A neural model of harmonic interference cancellation, *J. Acoust. Soc. Am.* **101**:2857-2865.

de Cheveigné, A., 1998, Cancellation model of pitch perception, *J. Acoust. Soc. Am.* **103**:1261-1271.

de Cheveigné, A., 1999a, Pitch shifts of mistuned partials: a time-domain model, *J. Acoust. Soc. Am.* **106**:887-897.

de Cheveigné, A., 1999b, Scale tree update, *ISO/IEC JTC1/SC29/WG11*, MPEG99/m5443.

de Cheveigné, A. and Kawahara, H., 1999, Multiple period estimation and pitch perception model, *Speech Communication* **27**:175-185.

de Cheveigné, A., 2001, Correlation Network model of auditory processing, *Proc. Workshop on Consistent & Reliable Acoustic Cues for Sound Analysis*, Aalborg (Danmark)

de Cheveigné, A., 2002, Scalable metadata for search, sonification and display, *Proc. International Conference on Auditory Display (ICAD 2002)*, 279-284.

de Cheveigné, A. and Kawahara, H., 2002, YIN, a fundamental frequency estimator for speech and music, *J. Acoust. Soc. Am.* **111**:1917-1930.

de Cheveigné, A. and Baskind, A., 2003, F0 estimation of one or several voices, *Proc. Eurospeech*, 833-836.

Cooke, M.P., 1991, *Modeling Auditory Processing and Organisation*, Sheffield, Department of Computer Science, unpublished doctoral dissertation.

Cooke, M. and Ellis, D., 2001, The auditory organization of speech and other sources in listeners and computational models, *Speech Comm.* **35**:141-177.

Cooke, M., Morris, A., and Green, P., 1997, Missing data techniques for robust speech recognition, *Proc. ICASSP*, 863-866.

Culling, J.F. and Summerfield, Q., 1995, Perceptual segregation of concurrent speech sounds: absence of across-frequency grouping by common interaural delay, *J. Acoust. Soc. Am.* **98**:785-797.

Culling, J.F., Summerfield, Q., and Marshall, D.H., 1998, Dichotic pitches as illusions of binaural unmasking I: Huggin's pitch and the "Binaural Edge Pitch", *J. Acoust. Soc. Am.* **103**:3509-3526.

Durlach, N.I., 1963, Equalization and cancellation theory of binaural masking level differences, *J. Acoust. Soc. Am.* **35**:1206-1218.

Ellis, D., 1996, *Prediction-driven Computational Auditory Scene Analysis*, MIT, unpublished doctoral dissertation.

Gábor, D., 1947, Acoustical quanta and the theory of hearing, *Nature* **159**, 591-594.

Hartmann, W.M. and Doty, S.L., 1996, On the pitches of the components of a complex tone, *J. Acoust. Soc. Am.* **99**:567-578.

ISO/IEC JTC 1/SC 29, 2001, *Information Technology—Multimedia Content Description Interface —Part 4: Audio*, ISO/IEC FDIS 15938-4.

von Helmholtz, H., 1877, *On the Sensations of Tone* (English translation A.J. Ellis, 1885, 1954), Dover, New York.

Hess, W., 1983, *Pitch Determination of Speech Signals*, Springer-Verlag, Berlin.

Jeffress, L.A., 1948, A place theory of sound localization, *J. Comp. Physiol. Psychol.* **41**:35-39.

Lippmann, R.P. and Carlson, B.A., 1997, Using missing feature theory to actively select features for robust speech recognition with interruptions, filtering, and noise, *Proc. ESCA Eurospeech*, KN-37-40.

Lyon, R.F., 1983, A computational model of binaural localization and separation, reprinted (1988) in *Natural Computation*, W. Richards, ed., MIT Press, Cambridge, Mass, pp. 319-327.

Morris, A.C., Cooke, M.P., and Green, P.D., 1998, Some solutions to the missing feature problem in data classification, with application to noise robust ASR, *Proc. ICASSP*, 737-740.

Nakamura, S. and Heracleous, P., 2002, 3-D N-Best Search for simultaneous recognition of distant-talking speech of multiple talkers, *Proc. IEEE ICMI.*

Palomäki, K., Brown, G.J., and Wang, D., 2001, A binaural model for missing data speech recognition in noisy and reverberant conditions, *Proc. CRAC (Consistent and Reliable Acoustic Cues) workshop*, Aalborg, Danmark.

Roweis, S., 2000, One-microphone source separation, in *Advances in NIPS*, Edited by M. Press, Cambridge MA, 609–616.

Roweis, S., 2003, Factorial models and refiltering for speech separation and denoising, *Proc. Eurospeech.*

Weintraub, M., 1985, *A Theory and Computational Model of Auditory Monaural Sound Separation*, Stanford unpublished doctoral dissertation.

Wu, M., Wang, D., and Brown, G. J., 2002, A multipitch tracking algorithm for noisy speech, *IEEE Trans. ASSP* **11**:229-241.

Chapter 17

Informational and Energetic Masking Effects in Multitalker Speech Perception

Douglas S. Brungart
Air Force Research Laboratory
douglas.brungart@wpafb.af.mil

1   INTRODUCTION

When a speech signal is obscured by a second simultaneous competing speech signal, two types of masking contribute to overall performance. Traditional "energetic" masking occurs when both utterances contain energy in the same critical bands at the same time and portions of one or both of the speech signals are rendered inaudible at the periphery. Higher-level "informational masking" occurs when the signal and masker are both audible but the listener is unable to disentangle the elements of the target signal from a similar-sounding distracter. Because "informational masking" is restricted to cases where the masking signal is similar to the target signal, it has a much greater impact on performance when a speech signal is masked by speech than it does when a speech signal is masked by noise. Furthermore, its effects depend specifically on the characteristics of the target and masking speech signals. This brief chapter outlines the results of some recent experiments we have conducted in our laboratory that have examined the role that informational masking plays in speech perception and attempted to isolate the effects that informational and/or energetic masking have on multitalker listening.

2   METHODS

All of the experiments described in this chapter were conducted using the Coordinate Response Measure (CRM). In the CRM task, a listener hears one or more simultaneous phrases of the form "Ready, (Call Sign), go to (color) (number) now" with one of eight call signs ("Baron," "Charlie," "Ringo," "Eagle," "Arrow," "Hopper," "Tiger," and "Laker"), one of four colors (red, blue, green, white), and one of eight numbers (1-8). Researchers at the Air

Force Research Laboratory have made a corpus of CRM speech materials available to the public on CD-ROM. This corpus contains all 256 possible CRM phrases (8 call signs X 4 colors X 8 numbers) spoken by each of eight different talkers (four male, four female). In the experiments described here, the stimulus always consisted of a combination of a target phrase, which was randomly selected from all of the phrases in the corpus with the call sign "Baron," and one or more masking phrases, which were randomly selected from the phrases in the corpus with different call signs, colors, and numbers than the target phrase. The listener's task was to listen for the phrase containing the preassigned target call sign "Baron" and respond with the color and number combination contained in that phrase. These stimuli were presented over headphones at a comfortable listening level (approximately 70 dB SPL), and the listener's responses were collected either by using the computer mouse to select the appropriately colored number from a matrix of colored numbers on the CRT or by pressing an appropriately marked key on a standard computer keyboard.

3   FACTORS THAT INFLUENCE INFORMA-TIONAL AND ENERGETIC MASKING IN SPEECH PERCEPTION

Figure 17.1 shows performance in the CRM listening task with five different maskers: speech-spectrum-shaped noise that has been amplitude-modulated to match the intensity fluctuations that occur in normal speech (TM); continuous speech-spectrum-shaped noise (TN); and a different-sex, same-sex, and same-talker speech signals (TD, TS and TT, respectively). The results shown in this figure highlight three important characteristics of informational masking in speech perception:

1. *The difference between speech-in-noise and speech-on-speech masking;* The two noise conditions shown in Figure 17.1 (TM and TN) are fundamentally different from the speech conditions in two important ways. First, performance with the noise maskers tends to remain at a high level at much lower SNR levels than performance with the speech maskers. Second, once the SNR does become low enough to degrade performance with the noise maskers, performance degrades monotonically and precipitously as the SNR is further reduced. In contrast, performance with the speech maskers (TD, TS, and TT) starts to degrade at much higher SNRs but degrades much more gradually, especially at negative SNR values.

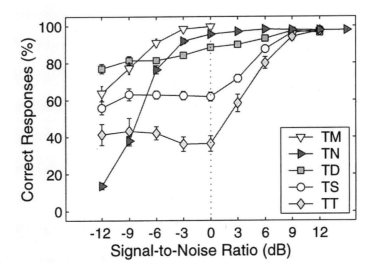

*Figure 17.1.*Color and number identifications as a function of signal-to-noise ratio for five types of masking signals: TM- envelope modulated speech-shaped noise; TN-continuous speech shaped noise; TD- a different-sex masking phrase from the CRM corpus; TS- a same-sex masking phrase from the CRM corpus; and TT- a masking phrase from the CRM corpus spoken by the same talker used in the target phrase. Adapted from Brungart (2001).

2. *The importance of voice characteristics:* Performance in the CRM task is much better with a different-sex interfering talker (TD) than with a same-sex interfering talker (TS), and much better with a same-sex interfering talker than with a masking phrase spoken by the same talker used in the target phrase (TT). Because informational masking depends on the relative similarity of the target and masking voices, differences in voice characteristics can be a powerful cue for segregating the target and masking speech signals.

3. *The advantages of level differences:* In contrast to performance with a noise masker, which degrades monotonically as the SNR decreases, performance with a same-sex speech masker tends to plateau around 0 dB SNR. The reason for this plateau in performance is that listeners are able to use differences in the levels of the two talkers to distinguish the two competing voices and selectively attend to the quieter of the two talkers in the stimulus. Thus, especially in the same talker (TT) condition, listeners may do better at negative SNR values because they can identify the target as the quieter talker in the stimu-

lus. In contrast, when the 0 dB SNR is 0 dB in the TT condition, the prosodic and coarticulative features that connect the call sign and color and number combination in the target phrase are the only available features to allow the listeners to discriminate between the color and number coordinates in target and masking voices.

Figure 17.2 shows how performance in the CRM listening task changes as additional masking talkers are added to the stimulus. When no competing talkers were present in the stimulus, performance was near 100%. The first competing talker reduced performance by a factor of approximately 0.4, to 62% correct responses. The second competing talker reduced performance by another factor of 0.4, to 38% correct responses. And the third competing talker reduced performance by another factor of 0.4, to 24% correct responses. Thus we see that CRM performance in a diotic multitalker speech display decreases by approximately 40% for each additional same-sex talker added to the stimulus.

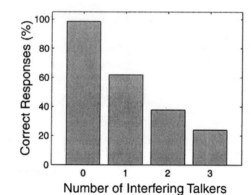

Figure 17.2. Performance in a diotic CRM listening task with 0, 1, 2, or 3 interfering same-sex talkers.

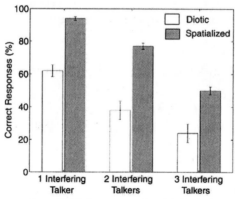

Figure 17.3. Performance in a CRM listening task with 0, 1, 2, or 3 interfering same-sex talkers, presented diotically or spatially separated by 45 degrees.

In general, informational masking is reduced whenever the attributes of the competing talkers are made more distinct in one or more perceptual dimensions. One very powerful way to distinguish the competing talkers in a multitalker stimulus is to spatially separate the apparent locations of the competing talkers. Figure 17.3 shows performance in the CRM task with 1, 2, or 3 competing talkers both in the diotic condition, where the talkers were presented from the same location, and in a spatial condition, where the talkers were spatially separated 45 degrees apart in azimuth. In the case with one interfering talker, spatial separation increased performance by approximately 25 percentage points. In the cases with two or three interfering talkers, spatial separation nearly doubled the percentage of correct responses. These results clearly illustrate the substantial decreases in informational masking that spatial separation in azimuth can produce in multitalker listening.

Figure 17.4 shows a final example of purely informational masking in dichotic speech perception. In this experiment, the normal two-talker same sex (TS) CRM speech stimulus was presented to the right ear. However, in this case, an additional speech noise masker was presented to the left ear (as indicated in the legend). The listeners were instructed to ignore the left ear and focus only on the right ear. The results show that a speech signal in the left ear interfered substantially with performance even when it was presented at a level 15 dB below the level of the target talker in the right ear, but that a noise signal in the left ear did not interfere even when it was presented at a level 20 dB louder than the target speech signal. In this case, the interference that occurred in the contralateral speech conditions was purely informational and had no energetic component. Ongoing research in our laboratory is now attempting to find other ways to isolate the informational and energetic components of speech on speech masking. Our hope is that this will result in a more complete understanding of the informational masking that occurs in speech and, in the long term, a significant improvement both in the audio displays that are used for multichannel speech communications and in the ability of automatic speech processing systems to process multitalker speech signals.

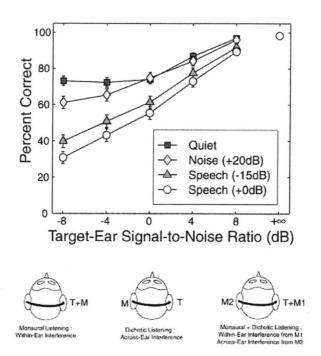

Figure 17.4. Performance in a dichotic CRM listening task with the target and one same-sex talker presented in the right ear and a masking signal (indicated by the legend) presented in the left ear.

References

Brungart, D.S. and Simpson, B.D., 2003, Within-ear and across-ear interference in a cocktail-party listening task: effects of masker uncertainty. In press, *Journal of the Acoustical Society of America.*

Brungart, D.S. and Simpson, B.D., 2002, Within-ear and across-ear interference in a cocktail-party listening task. *Journal of the Acoustical Society of America,* **112(6)**, 2985-2905.

Brungart, D.S. and Simpson, B.D., 2002, The effects of spatial separation in distance on the informational and energetic masking of a nearby speech signal. *Journal of the Acoustical Society of America,* **112(2)**, 664-676.

Brungart, D.S., Simpson, B.D., Scott, K.R., and Ericson, M.A., 2001, Informational and energetic masking effects in the perception of multiple simultaneous talkers. *Journal of the Acoustical Society of America,* **110(5)**, 2527-2538.

Brungart, D.S., 2001, Evaluation of speech intelligibility with the Coordinate Response Measure. *Journal of the Acoustical Society of America.* **109(5)**, 2276-2279

Brungart, D.S., 2001, Informational and energetic masking effects in the perception of two simultaneous talkers. *Journal of the Acoustical Society of America,* **109(3)**, 1101-1109.

Darwin, C.J., Brungart, D.S., and Simpson, B.D., 2003, Effects of fundamental frequency and vocal-tract length changes on attention to one of two simultaneous talkers. *Journal of the Acoustical Society of America,* **114(5)**, 2913-2922.

Ericson, M.A., Brungart, D.S., and Simpson, B.D., 2003, Factors that influence intelligibility in multitalker speech displays. In press, *International Journal of Aviation Psychology.*

Kidd, G., Mason, C.R., Arbogast, T.L., Brungart, D.S., and Simpson, B.D. (2003) Informational masking caused by contralateral stimulation. *Journal of the Acoustical Society of America,* **113(3)** 1594-1603.

Chapter 18

Masking the Feature Information In Multi-stream Speech-analogue Displays

Pierre L. Divenyi
Veterans Affairs Medical Center and
East Bay Institute for Research and Education
Martinez, California
pdivenyi@ebire.org

1   INTRODUCTION

Separation of speech signals by humans is one of the auditory-nervous processes we all count on to occur automatically and efficiently. Fortunately, this is mostly the case, as long as the individual has no, or at most mild, hearing loss and he/she is relatively young. This phenomenon has been termed "the cocktail-party effect" (CPE) by Cherry (1953). Deficient speech separation – a CPE deficit – is observed in people with moderate-to-severe sensorineural hearing loss regardless of age (Souza and Turner 1994), but also in elderly individuals regardless of their peripheral hearing sensitivity and regardless of whether or not the target and the (babble-) masker speech sources are spatially separated (Gelfand *et al.*, 1988; Divenyi and Haupt 1997). During the last 25 years, much effort has been devoted to investigate the CPE, its characteristics, its component and underlying processes, its failures in certain class of listeners—especially the elderly—as well as computational models emulating or superseding the human biological system that allows it to take place. To this date, however, we still have not reached the point at which the processes responsible for CPE and the causes of its dysfunctions would be fully understood, or at which computational means of separating simultaneous speech signals would be reliable to a degree permitting the machine to take over when humans fail or when they are not even present.

Previous work on the perceptual separation of simultaneous speech sources in our laboratory has been aimed at trying to understand the mechanisms that play a role in the CPE in the young, and in its decline in the elderly. The focus of much of this work has been identification of the significant dimensions of the phenomenon, such as perceptual segregation of sources—

"streams," as they are called following the terminology by Bregman (1990)—based on differences with regard to spatial location, fundamental-frequency pitch, formant frequency, and/or syllabic (or subsyllabic) rhythm. For these studies, we used simultaneous pairs of brief signals (typically shorter than 500 ms) that retained only very basic characteristics of speech: f_0, a single formant, temporal envelope pattern. One surprising finding of this research was that spatial separation of two streams provided only moderate advantage even for experienced young listeners in a stream segregation task (Divenyi 2001). This finding places a heavy emphasis on speech separation performed without the benefit of spatial cues, i.e., in a single spatial channel—as when listening to an amalgam of voices through a single loudspeaker. To successfully perform this task, the listener must rely on other cues, e.g., pitch difference, spectral pattern difference, and temporal pattern difference between the concurrent speech streams.

But how do simultaneous streams of speech interfere with one another? For the situation of a single target speaker's speech embedded in the babble of six to twelve other speakers, one is tempted to attribute interference to masking. However, since interference also occurs by the presence of a single unwanted source that allows the target to remain audible at least part of the time and in part of the spectrum, traditional, or *energetic*, masking has been shown to account for not more than part of this interference. As Brungart describes in chapter 17 in the present book, much of the masking in these situations is *informational*: the target may be above energetic-masked threshold but the information therein is fully or partially blocked by the presence of similar, although not identical, information in the interferer. One reason for the existence of informational, in addition to energetic, masking in speech is that, according to its most widely accepted definition, energetic masking implies stationary signals and maskers. Speech, on the other hand, is quintessentially dynamic, characterize by Plomp as a signal "slowly varying in amplitude and frequency" (Plomp 1983). Thus, to uncover the nature of speech-by-speech interference, it will be necessary to look at its *dynamic features*. By *features*, we mean spectro-temporal acoustic patterns, that may or may not coincide with phonetic or phonological features, and by *dynamic* we mean that these patterns are undergoing slow (2 to 20-Hz) amplitude modulation (AM) and/or frequency modulation (FM).

In our investigations of properties of the interference, we first decided to focus our attention on only one given feature in the target stream in the presence of interference by the same feature in another – the distractor stream. The specific question we asked was: how resistant are we to interference when identifying a pattern of slow AM or FM fluctuations? The question cast in psychophysically tractable terms is: in a target stream containing a pattern of a given feature, what degree of informational masking would be produced by a simultaneously present distractor stream that contains a random sequence of the same feature?

In the listening experiments described in the following paragraphs, two such features were investigated: syllabic-rate rhythmic pattern (AM) and formant excursion pattern (FM).

2 EXPERIMENT 1 – METHODS

In order to answer the general question, both the problem and the signals had to be stripped down to their most essential characteristics. The speech target became a harmonic complex tone of a given fundamental frequency f_0, with the frequency of its components limited between 400 and 3,300 Hz. The interfering distractor was another harmonic complex with components between the same spectral limits but an f_0 clearly different from that of the target (either by a factor of 0.27, i.e., close to a Major third in music, or a factor of 0.77, close to a minor seventh). The lower of the two fundamental frequencies was always 107 Hz, i.e., an f_0 corresponding to an average male voice.

Experiment 1 addressed the question of the way listeners identify a slow rhythmic target pattern consisting of AM bursts in the presence of a distractor consisting of similar AM bursts but separated by random time intervals. The objective of the experiment was to assess quantitatively informational masking of a syllabic rhythmic pattern in one stream by a sequence of random, arrhythmic bursts in another stream. The target pattern was either a "o-o—o" amphibrach or a "o—o-o" dactyl, differing from a rhythmically regular "o-o-o" spondee by 10 to 50 percent, depending on the condition. The carrier signals of the target and the distractor were the sinusoidal complexes described in the preceding paragraph. Informational masking was assessed in one of two ways. In experiment 1a, as illustrated in Figure 18.1, the degree of irregularity of the pattern and the absolute level of the distractor were fixed; the presentation level of the distractor was 87 dB SPL while that of the target was adaptively changed, in a three-down-one-up two-alternative forced-choice paradigm, to track the target level yielding a 79.4 percent performance level (Levitt 1971). The interval separating the onsets of the first and the last (=third) pattern burst was 800 ms, i.e., the average duration between consecutive burst onsets was 400 ms. In experiment 1b, the levels of the target and the distractor were held constant (80 dB and 74 dB SPL, respectively) and it was the degree of rhythmic irregularity of the pattern that was adaptively varied.

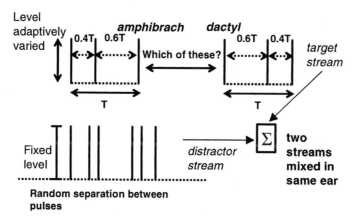

Figure 18.1. Schematic diagram of the stimulus of experiment 1a. The target pattern stream, an amphibrach or a dactyl, is shown in the top row and the distractor stream in the bottom row. In the condition shown, each of the two patterns departs from a regular spondee rhythm by 20% (=0.1/0.5). Each vertical pulse represents a burst of complex sinusoid 40 ms in duration. The two streams are presented mixed in the same spatial channel. The listener's task is to identify the target pattern in the random distractor. The level of the target is adaptively varied from trial to trial to track the level at which identification is 79.4% correct.

In this sub-experiment, the average time interval between consecutive pattern bursts was 100, 200, or 400 ms. Throughout Experiment 1, whether the target was assigned the higher f_0 and the distractor the lower, or vice versa, was varied from condition to condition.

Six young adults with normal hearing (average age=22±3.4 yrs) and seven elderly individuals (average age=68.7±7.1 yrs) with mild-to-moderate hearing loss served as paid listeners in the experiment. Hearing loss of the elderly subjects was 27.8±19.8 dB at 4 kHz; their mean pure-tone average of four frequencies between 500 and 4 000 Hz was 20.6±14.2 dB. The young subjects had extensive experience (6 months or more) as subjects in psychoacoustic experiments. The elderly subjects had at least one month of experience in similar experiments. The subjects were seated in a sound-attenuated room and listened to runs of stimuli under diotically wired earphones. One run consisted of one adaptive threshold determination. Trial-by-trial visual feedback was provided.

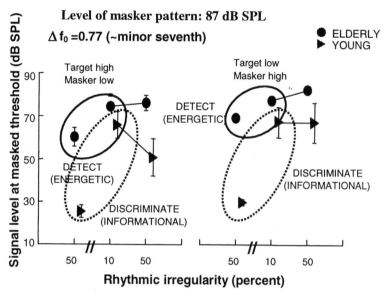

Figure 18.2. Masked thresholds obtained in experiment 1a for the elderly and the young subjects. The data clearly show that informational masked thresholds are higher than energetic masked thresholds: it is easier to detect than to discriminate the rhythmic patterns in the target. Having the target in the higher- and the distractor in the lower-f_0 stream yields somewhat lower masked thresholds than the opposite. Finally, masked thresholds — both energetic and informational — increase with age.

2.1   Experiment 1 – Results

Results for the elderly and young listeners are illustrated in Figure 18.2 for experiment 1a, in which the level of masked thresholds (both energetic and informational) of the rhythmic patterns was assessed. The fundamental frequency difference between target and distractor streams was just a little shy of one octave; in other words, the pitch of the higher stream was comparable to that of a woman's and that of the lower to a man's voice. Thresholds for informational masking were higher than for energetic masking — especially for the younger subjects. This difference is mainly attributable for inorderly high energetic masking thresholds obtained for elderly listeners, which is likely to be the result of difficulty by these subjects to perceptually segregate the target from the distractor stream. In other words, the elderly don't seem to be able to correctly assign a given burst to the correct streak (target or distractor).

But informational masking, in addition to requiring a higher intensity for the target in order to be discriminated, can also influence the degree of irregularity on the basis of which discrimination of the two target patterns (amphibrach and dactyl) can be performed. The degree of irregularity necessary to achieve threshold performance was assessed in experiment 1b, the results of which are shown in the two panels of Figure 18.3 separately for the elderly (left panel) and for the young (right panel) subjects. Looking at the young subjects' data, it becomes apparent that their discrimination of rhythmic irregularity approaches the limits of auditory time discrimination (about 6 percent, see Divenyi and Danner 1977): for all conditions except one (target in low stream, target-distractor stream f_0 difference small [=0.27], sequence rapid [average separation of 100 ms between consecutive bursts]), at which irregularity discrimination reached 15 percent, irregularities between 4 and 8 percent could be reliably discriminated. In contrast, the condition most difficult for the young to discriminate by the young could not be discriminated by the elderly at any irregularity, no matter how large. The irregularity thresholds obtained for the elderly were at best between 15 and 20 percent but increased to 42 to 50 percent when the sequences were rapid. Just as in Experiment 1a, target pattern discrimination was generally better when the f_0 of the target was higher than that of the distractor, and when the fundamental frequency separation was relatively large.

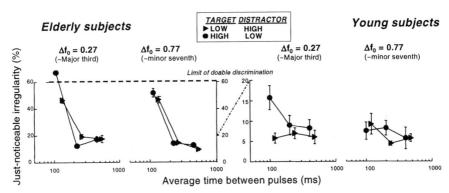

Figure 18.3. Results of Experiment 1b: percent irregularity at the 79.4 percent discriminability threshold (on the ordinate) obtained for the elderly (left panel) and the young (right panel) subjects. Note that the ordinate scale for the young is 1/3 of the scale for the elderly. The abscissa displays the average time interval between consecutive bursts. Results for both narrow (0.27) and wide (0.77) fundamental frequency separation between target and distractor are illustrated for the target having the high f_0 and the distractor the low f_0, as well as the opposite. Discrimination of irregularity larger than 60 percent is not doable because two of the bursts become too close for the subject to hear them as two for the most rapid rhythmic sequence condition. The target and distractor levels were held constant at 80 and 74 dB SPL, respectively.

3   EXPERIMENT 2 – METHODS

Perceptual segregation of pairs of simultaneous patterns differing in fundamental frequencies and formant values is easier when the formant frequencies dynamically change (Divenyi *et al.*, 1997). The objective of experiment 2 was to measure informational masking of slowly changing FM pattern imposed on a single formant in a complex sinusoidal carrier, by a distractor consisting of continuously up-down changing formant frequency, as illustrated in Figure 18.4. The distractor's formant frequency excursion was twice that of the target pattern's formant frequency excursion and was started at a phase that randomly changed from trial to trial. The duration of the upward and downward formant transitions was 100 ms in both the target and the distractor. In the target, the up-down and the down-up glide patterns were preceded and followed by a steady-state portion 100 ms in duration. The fundamental frequency of the target and that of the distractor were different and had the same values as those used in experiment 1. Again, the target could be presented at the higher f_0 and the distractor at the lower, or vice versa, depending on the experimental condition. The subject's task was to identify the formant glide pattern as either up-down or down-up. The intensity level of the distractor was held constant at 67 dB SPL and that of the target was varied adaptively to track the one at which the performance reached 79.4 percent correct in a two-alternative forced-choice paradigm. In experiment 2a energetic and informational masking of the target by the distractor were compared for a constant 60 percent target formant frequency excursion. In experiment 2b only informational masking was measured for formant frequency excursions ranging from 10 to 60 percent.

3.1   Experiment 2 – Results

Unsurprisingly, as illustrated in Figure 18.5, energetic masking of the formant glide pattern by a 67 dB SPL distractor was very ineffective: up to distractor levels of close to 60 dB SPL, target patterns are detectable near their threshold in quiet (about 15 dB SPL). This happens because the spectrotemporal patterns of the two signals being quite different, the target and the masker slip in and out spectral regions with masker energy large enough to cause masking. Also, very similarly to what we observed in experiment 1a, informational masking thresholds are much more elevated than energetic masking thresholds: by between 12 and 19 dB for young and by as much as 35 to 40 dB for elderly listeners.

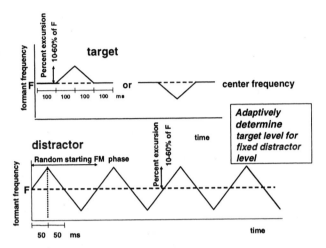

Figure 18.4. Schematic spectrogram-type diagram of the stimulus in experiment 2a. The lines must be understood as representing peaks of a single resonance imposed on a complex sinusoidal carrier having components between 400 and 3300 Hz but different fundamental frequencies f_0 for the target pattern and for the distractor. The level of the distractor was constant at 67 dB SPL, whereas that of the target pattern was varied adaptively to track one at which the target pattern could be identified 79.4 percent at a time. The frequency swing, or excursion, of the glide was constant at 60 percent. The stimulus in experiment 2b was similar, except that a series of different frequency excursion values between 10 and 60 percent was explored.

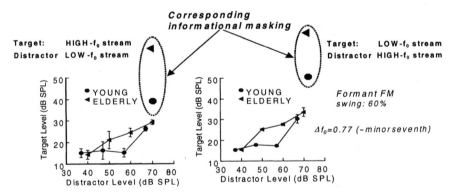

Figure 18.5. Results of Experiment 2a: Energetic masking corresponding to the 79.4 percent detection threshold (on the ordinate) obtained for the elderly and the young subjects for the target in the high-f_0 stream and the distractor in the low-f_0 stream (left panel) and the opposite (right panel). The abscissa displays the level of the distractor. Results are for the wide (0.77) fundamental frequency separation between target and distractor and for the 60 percent formant excursion only. The two data point pairs circled indicate, for the comparison's sake, informational masking thresholds for the 67-dB SPL distractors.

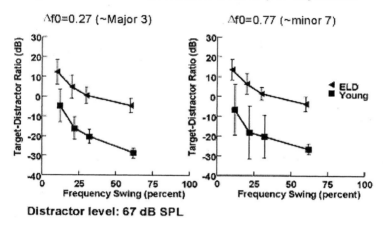

Distractor level: 67 dB SPL

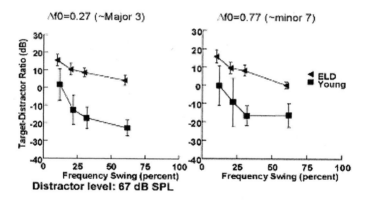

Figure 18.6. Results of Experiment 2b: informational masking corresponding to the 79.4 percent discrimination threshold (on the ordinate) obtained for the elderly and the young subjects with the target in the high-f_0 stream and the distractor in the low-f_0 stream (the two graphs on the left) and the opposite (the two graphs on the right). The abscissa displays formant frequency swing (excursion) expressed in percent of the starting and ending formant frequency. Results are shown both for the narrow (0.27) and the wide (0.77) fundamental frequency separation between target and distractor.

Again, targets in the low f_0 stream are more prone to be masked than targets in the high f_0 stream. Thus, results of experiments 1a and 2a are internally consistent.

Informational masking as a function of formant glide excursion was examined in experiment 2b with the results illustrated in Figure 18.6. The extent of formant excursion is inversely related to the amount of informational masking, as expected. However, the exact numbers are surprising. For one, young listeners are able to perform the glide pattern discrimination task with glide intensities at or below the intensity of the distractor, even when the pattern and distractor fundamental frequencies are close to each other and the swing is only 10 percent. At large frequency swing values, they can discriminate the glide patterns even at intensities 20 to 30 dB below the one of the distractor. Although the formant swing-informational masking functions obtained for the elderly listeners are very similar to those of the young, the are shifted upward by about 15 dB in the difficult conditions to as much as 25 to 30 dB in some of the easier conditions, indicating that perceptual separation of a target glide pattern from a distractor consisting of continuous up-down formant transitions becomes impaired in aging even when the pattern and the distractor are presented well above audibility thresholds.

4   DISCUSSION

We have examined perception of a speech analog target having a given fundamental frequency, presented simultaneously with a distractor having a different fundamental frequency. In the first experiment, the target was a slow-AM burst pattern (mimicking syllabic fluctuation) and the distractor a random sequence of similar bursts. In the second experiment, the target was a slow-FM formant glide pattern (mimicking vowel-to-vowel transitions) and the distractor several cycles of FM formant glides presented at a random starting FM phase. First, we saw that, at low target-to-distractor (T/D) ratios, the detectability of the target was masked. We termed the T/D ratio at which the target was just detectable the threshold of *energetic masking*. However, in order for the listeners to be able to discriminate the target pattern, the T/D ratio had to be increased. Since in those cases the interference by the distractor affected the perception of *information* carried by a given dimension of the target (envelope fluctuation or formant trajectory) rather than its *audibility*, we termed the T/D ratio at which the target pattern was just discriminable the threshold of *informational masking*. The basic distinction between the two types of masking is that the masker in informational masking, i.e., the distractor, has information that the subjects easily confuse with similar information in the target. In other words, masking takes place because the target stream and the distractor stream are being *imperfectly segregated* by the listener. As opposed to energetic masking that suppresses detectability by way of overlapping spectrotemporal fields between the target and the distractor, informational masking is a process that cannot be explained by peripheral interference. Attributing the interference seen in the present informational masking results to central processes is in agreement with conclusions by other investigators (e.g., Kidd *et al.*, *et al.* 1998; Oh and Lutfi 1999; Brungart *et al.*, 2001), albeit our definition of the difference between energetic and informational masking is not strictly identical to that found in the literature.

We saw that for normal-hearing young listeners, the T/D ratio at which the target AM or FM patterns can be identified is surprisingly low: typically -30 to -15 dB depending on the condition, i.e., somewhat higher when the target and distractor streams are close in fundamental frequency or when the AM fluctuation is rapid or the FM fluctuation depth is low. However, the T/D ratio necessary for pattern identification dramatically increases, to between 0 and +10 dB, for elderly listeners not only with mild-to-moderate hearing loss but also those with no hearing loss at all. Since these listeners also exhibit deficits in the understanding of speech in "cocktail-party" situations, such as amidst babble noise or in reverberant environments, the informational masking deficit and the CPE deficit reflects a parallel or could possibly signify a causal relationship between the underlying processes. Age effects observed for other types of informational masking (e.g., Kidd *et al.*, 2002) may also be related to the same deficits.

What do these results mean for speech separation in general? First, we want to point out that the two types of targets (and distractors) investigated represent two among the *dynamic processes* that speech is built on. As we said earlier, speech is defined as a process slowly varying in amplitude and frequency; the AM and FM changes in the target, although oversimplified when compared to real speech, address characteristics that are most important in speech as well as speech separation. When two speech streams, each independently varying along the amplitude and frequency dimensions, are presented simultaneously, their mutual interference will also change in time, just as two spectrograms plotted on top of one another will have spectrotemporal areas where speech can be "glimpsed" at – as described by Cooke in chapter 21 in the present book. Our results help specify some of the parametric constraints under which such "glimpsing" will be successful, and others under which it will fail. Generally speaking, the negative signal-to-noise ratios at which target identification remains successful indicates that the dynamic changes themselves may serve as pointers (in spectrum and time) that help the perceptual apparatus find the missing spectrotemporal information when it is covered by an interfering signal. As such, the present findings could provide one piece of suggestion for improving computational auditory scene analysis (CASA) results: the direction of change in amplitude and/or frequency could be used to recover missing data.

While the above suggestion puts a definitely positive spin on the present results, we cannot ignore the unfortunate fact that the aging process puts a serious dent into the human ability to separate dynamically changing signals. Although no known remedy exists for this dysfunction, it is our hope that computational schemes can be applied for the design of a device that will help re-integrate in Society the vast elderly population suffering from CPE disorder.

5   ACKNOWLEDGMENT

The author wishes to thank Brian Gygi and Alex Brandmeyer for their help during the experimental stage of the research. The research has been supported by grant R01-AG07998 from National Institute on Aging and by the VA Medical Research.

References

Bregman, A.S., 1990, *Auditory scene analysis: The perceptual organization of sound.* Bradford Books (MIT Press), Cambridge, Mass.

Brungart, D.S., Simpson, B.D., Ericson, M.A., and Scott, K.R., 2001, Informational and energetic masking effects in the perception of multiple simultaneous talkers. *J Acoust Soc Am,* **110(5 Pt 1)**, 2527-2538.

Cherry, C., 1953, Some experiments on the recognition of speech with one and with two ears. *Journal of the Acoustical Society of America,* **26**, 975-979.

Divenyi, P.L., and Danner, W.F., 1977, Discrimination of time intervals marked by brief acoustic pulses of various intensities and spectra. *Perception and Psychophysics,* **21**, 125-142.

Divenyi, P.L., Carré, R., and Algazi, A.P., 1997, Auditory segregation of vowel-like sounds with static and dynamic spectral properties. In *IEEE Mohonk Mountain Workshop on Applications of Signal Processing to Audio and Acoustics,* D. P. W. Ellis, ed., IEEE, New Paltz, N.Y., pp. 14.11.11-14.

Divenyi, P.L., and Haupt, K.M., 1997, Audiological correlates of speech understanding deficits in elderly listeners with mild-to-moderate hearing loss. I. Age and laterality effects. *Ear and Hearing,* **18(1)**, 42-61.

Divenyi, P.L., 2001, Dimensions of auditory segregation: What do they tell us about levels of auditory processing? In *Physiological and Psychophysical Bases of Auditory Function,* D.J. Breebart, A.J.M. Houtsma, A. Kohlrausch, V.F. Prijs, and R. Schoonhoven, eds., Shaker Publishing BV, Maastricht, the Netherlands, pp. 468-476.

Gelfand, S.A., Ross, L., and Miller, S., 1988, Sentence reception in noise from one versus two sources: Effect of aging and hearing loss. *Journal of the Acoustical Society of America,* **83**, 248-256.

Kidd, G., Jr., Mason, C.R., Rohtla, T.L., and Deliwala, P.S., 1998, Release from masking due to spatial separation of sources in the identification of nonspeech auditory patterns. *J Acoust Soc Am,* **104(1),** 422-431.

Kidd, G., Jr., Arbogast, T.L., Mason, C.R., and Walsh, M., 2002, Informational masking in listeners with sensorineural hearing loss. *J Assoc Res Otolaryngol,* **3(2),** 107-119.

Levitt, H., 1971, Transformed up-down methods in psychoacoustics. *Journal of the Acoustical Society of America,* **49,** 467-477.

Oh, E.L., and Lutfi, R.A., 1999, Informational masking by everyday sounds. *J Acoust Soc Am,* **106(6),** 3521-3528.

Plomp, R., 1983, Perception of speech as a modulated signal. In *Proceedings of the Tenth International Congress of Phonetic Sciences,* M.P.R.v.d. Broecke, and A. Cohen eds., Dordrecht, Foris., pp. 29-40

Souza, P.E., and Turner, C.W., 1994, Masking of speech in young and elderly listeners with hearing loss. *Journal of Speech and Hearing Research,* **37,** 665-661.

Chapter 19

Interplay Between Visual and Audio Scene Analysis

Ziyou Xiong
Dept. of Computer And Electrical Engineering
University Of Illinois At Urbana-champagne
zxiong@ifp.uiuc.edu

Thomas S. Huang
Dept. of Computer And Electrical Engineering
University Of Illinois At Urbana-champagne
huang@ifp.uiuc.edu

1   INTRODUCTION

Computational auditory scene analysis (CASA) (Cooke, 1993) and visual scene analysis have been researched from two separate communities, namely CASA and Computer Vision. Separation of audio sources can be very useful for visual scene analysis. Some visual scene analysis techniques can also be used to help audio separation. Recently many machine learning algorithms (Jojic and Frey, 2001; Tao *et al.*, 2000; Wilson and Bobick, 1999; Frey and Jojic, 1999; Jojic *et al.*, 2000; Williams and Titsias, 2002; Frey and Jojic, 2003) have been shown to be useful in video tracking, video analysis and understanding. We have begun to apply these algorithms designed originally for visual scene analysis to audio scene analysis. We report some results here. The joint audio-visual scene analysis will potentially open more venues for further research.

2   AUDIO SCENE ANALYSIS HELPS VISUAL SCENE ANALYSIS

Tasks like video indexing, summarization and retrieval usually start with low-level feature extraction followed by analysis based on these features or based on modeling "high-level" concepts using these features. Examples of these features are color histogram, texture and shape. Examples of the "high-level" concepts are play or break in soccer games, in-door or out-door scenes.

The accuracy of these tasks can be greatly improved when scene analysis is also carried out in the audio. There are two reasons for this integration. First, many "high-level" concepts such as explosion, rocket launching are audio-visual events. They are better modeled using joint audio-visual cues. Next, some "high-level" concepts are more easily modeled in the audio domain than in the video domain. An example is the detection of humorous moments in sit-coms. The detection of audience laughter is easier than the detection of humorous video gestures or actions. We have shown that audio analysis alone can achieve good results in tasks such as sports highlights extraction. We have also shown that a joint analysis from both two domains out-perform either of the single-domain analysis (Xiong *et al.*, 2004).

There are several research issues on how audio scene analysis helps visual scene analysis:

- **Recognition without audio source separation.** In the example of sports highlights extraction, the audio sound track has a lot of audio mixtures such as applause with commentator's speech, music with speech in the commercial sessions. In the latter case, although it is a mixture of speech with music, the separation will not be necessary for the task of deciding whether it is highlight or not.

- **Separation then recognition.** In the above example of applause with commentator's speech, if we can separate the applause from the commentator's speech then the applause can be used to identify the highlight more accurately. Without separation, it might be difficult to recognize the applause sound in the mixture in the first place.

- **What is the best way to fuse the audio scene and visual scene analysis?** Many fusion techniques have been proposed in literature such as coupled Hidden Markov Models (HMMs), Dynamic Bayesian Nets (DBNs) etc. We have shown the advantage of coupled HMMs over single modality HMMs, but we are aware of the shortcomings in that system. These include the difficulty locating the onset and offset of the targeted events, the existence of unexpected events that are not included in the learned models of the pre-defined events and so on.

- **How to use the stereo sound to help audio scene analysis?** Broadcast video content usually comes with stereo audio signals. However, litter work has been done to take advantage of this fact. Note that this is different from separation of two speakers from two microphone recordings.

3   VISUAL SCENE ANALYSIS HELPS AUDIO SCENE ANALYSIS

The problem of audio source separation has its counterpart in visual scene analysis, from image foreground/background separation, image segmentation to event detection in video. So many of the techniques developed in image/ video domain can potentially be useful to separate audio sources.

Probabilistic machine learning techniques will be our main tools. Recently, we have found that *generative* probabilistic models (GPMs) are especially useful for visual scene analysis, and we have also obtained preliminary results using GPMs for microphone array analysis.

3.1   Probabilistic Models

Probabilistic models are natural candidates to use when analyzing complex data, such as video frames and audio signal, which contain many uncertainties, such as which object a given pixel belongs to. A complete description of a probabilistic model requires the joint probability of all random variables. Since the number of random variables can be very large, we run into problems when trying to estimate the full joint distribution from a limited amount of data; when trying to store the exponentially large number of parameters; and when trying to compute marginal probabilities, which take exponential amounts of time. Fortunately, in most scenarios, the random variables are not all interdependent. Each random variable is usually directly dependent only on a small subset of the remaining random variables (the Markov blanket). Dependencies on other variables are mediated through the variable's Markov blanket. Thus, the complete joint probability distribution (or density function) can be factored into simpler components. This factorization can be efficiently described using graphs, and in particular Bayesian Networks (BNs) (Pearl, 1988).

In statistical pattern analysis, there are two major categories of probability models: discriminative and generative. In the discriminative approach, we model the conditional probability of the class label given the observation and are usually not concerned with how to explain the observed data. A generative model, on the other hand, includes any hidden variables that are useful in accounting for structured effects, such as variations in illumination. In a generative model, if all the probabilities are known, then one can generate a sample observation from the model that will resemble real observed data. In analysis tasks, we are given the observed data and must find a probability model that reflects prior structural knowledge, and efficiently explains the data. Model estimation is often performed using the expectation-maximization (EM) algorithm and similar techniques (Dempster *et al.*, 1997).

3.2   Generative Probabilistic Model (GPM) in Visual Scene Analysis

Recently, GPM (specifically, BN) have been applied to the problem of automatic visual scene interpretation and object tracking (Tao *et al.*, 2000; Wilson and Bobick, 1999; Frey and Jojic, 1999; Jojic *et al.*, 2000; Williams and Titsias, 2002; Frey and Jojic, 2003). Our earliest work was to detect in an unsupervised way recurrent spatial patterns in a video sequence, for example, to find image clusters in the set of images in the central panel of Figure 19.1. Notice that there are two human subjects in this set of images and that their positions vary among the images. The traditional Expectation-Maximization (EM) algorithm takes these images and tries to fit a mixture of Gaussian probability distribution. The means of the learned mixtures are shown in the lower-right panel of Figure 19.1 for a case of 4 mixtures. Due to the fact that the observed images are not well aligned, these mean images are quite blurry. Frey and Jojic (2003) have introduced a GPM (see the leftmost panel of Figure 19.1) to model how those images are generated. First, an image class c is generated. Then depending on which class it is, a random variable z is generated, which is assumed to be Gaussian. Its mean is the mean of the images in class c. Next a random variable s is generated which specifies the spatial shift to be applied to z. Given s and z, an observed image x is generated. x is also a random variable with the mean being the shift version of z, i.e. $s(z)$. The inference problem is that given only the observed x, how to estimate s, z and c. Frey and Jojic addressed this problem in (2003). Their results on the same set of data are shown in the upper-right panel of Figure 19.1. These cluster centers are much less blurry.

The original algorithm in Frey and Jojic (2003) worked in the batch mode. More recently, we have developed an on-line version and have used it to cluster frames in videos (Jojic *et al.*, 2003).

To account for multiple objects that may occlude each other, a layered representation of the 3-D scene can be included in a hierarchical BN (Jojic *et al.*, 2000; Jojic *et al.*, 2003). This BN accounts for the locations, 2-D appearances and 2-D shapes of multiple, occluding objects in a 3-D scene. The total number of configurations of the object instantiation parameters is too large for exact inference, but an approximate variational method can be used to approximately compute probabilities and perform generalized EM learning (Neal and Hinton, 1998). An example of the decomposition of an input video frame into its layers after learning is shown in Figure 19.2. Note that the only input to the variational EM algorithm was the number of layers and the input video.

Figure 19.1. Left: The GPM to detect recurrent spatial patterns in a video sequence. Middle: A set of image examples. Upper-Right: Cluster centers found using the GPM method. Lower-Right: Cluster centers found by the traditional EM algorithm.

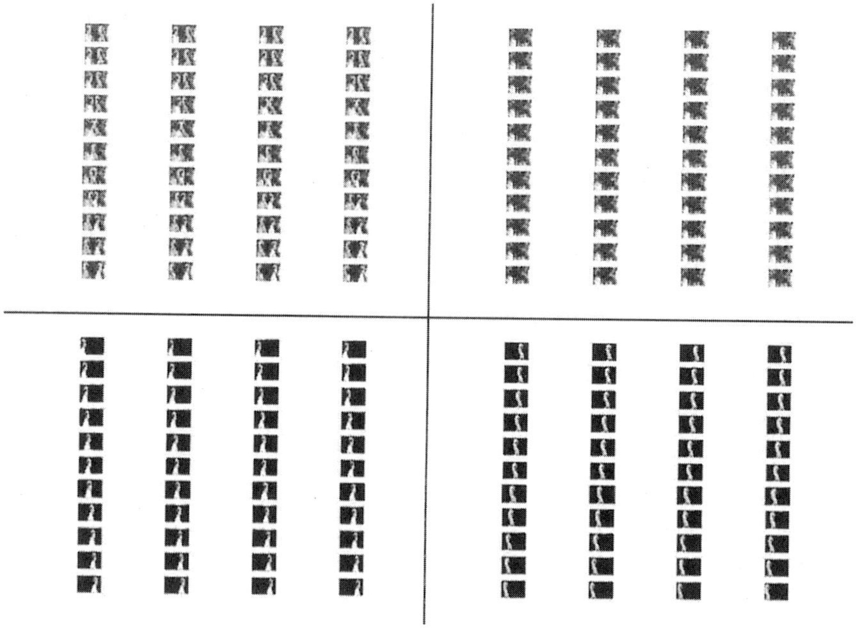

Figure 19.2. Three layers were learned from the input video (frames at the upper-left quadrant), and then probabilistic inference was used to re-render the video frames, automatically deriving various layers (background frames at the upper-right, frames of one person at the lower-left and frames of another person at the lower-right quadrant). Note that the positions of the two persons in the images are correctly inferred.

In this model, each object is described by a 2-D "cardboard cut-out", i.e., an appearance map and a transparency map (mask) that specifies whether each pixel is transparent or opaque, or somewhere in-between. Each cut-out can appear at any location in the scene and importantly. Jojic and Frey (2001) showed how a variational technique can be used for approximate probabilistic inference and learning in this model. After learning, probabilistic inference can be used to locate objects and their instantiation parameters (e.g., position of a person). Since the above technique makes use of a principled probability model, this model can be combined with other generative probabilistic models, such as models of audio, to perform joint multi-modal analysis tasks.

4   SEPARATION OF AUDITORY CORRELOGRAM

Inspired by the visual scene analysis algorithm in Jojic and Frey (2001), we explore the possibility of using existing GPMs (developed for visual scene analysis, such as layering) directly on suitably chosen representations of audio signals. In particular, we think a potentially useful representation is the audio correlogram (Slaney *et al.*, 1994).

The audio correlogram is a 3-D representation of an audio signal. It is derived by taking the short-time autocorrelation of each frequency band in a 2-D time-frequency representation of the audio signal:

$$A(i, j, \tau) = \sum_{k=0}^{K-1} r(i, j-k) r(i, j-k-\tau) w(k), \tag{1}$$

where i is the frequency index, j is the time index taken K samples at a time, τ is the autocorrelation delay index, r is the 2-D time-frequency representation of the audio signal which is chosen to be the audio cochleagram (Slaney *et al.*, 1994) and w is a short-time rectangular window. Slaney *et al.* (1994) have shown that it is possible to reconstruct the original audio signal from its audio correlogram almost perfectly.

We propose to develop an inference algorithm that can separate the correlogram of a sound mixture into correlograms of the individual sources, and reconstruct the sources using the separated correlograms. In the following, we describe this approach in the case of separating 2 sources from 1 mixture.

The observed pixel intensities x in a frame of the correlogram, i.e, $A(:, j, :)$, is assumed to be explained by combining a transformed version of a vector of pixel intensities s_1 of the first source and a transformed version of a vector of pixel intensities s_2 of the second source:

$$x = T_2 * s_2 + T_1 * s_1 + n, \tag{2}$$

where "*" indicates element-wise product. The transformation operators T_1 and T_2 account for the change of fundamental frequencies of the two sources respectively. Correlogram frames display an expanding-and-compression effect which relates to the up-and-down changing pattern of the fundamental frequency. This kind of change can be accounted for by a horizontal shift if the coordinate axis is measured on a log-scale.

Since each frame of the correlogram is the auto-correlation of the frequency response on the cochleagram, if the two audio sources are assumed to be uncorrelated, then it is a valid assumption that the correlogram of the mixture signal is the sum of the those of the two audio signals. This additive property is shown in Figure 19.3. The left-most column and the central-column have several examples of correlogram frames. The right-most column has the corresponding correlogram frames of the mixture audio signal. Since the frames of the correlogram are similar to frames in a video sequence, we believe that the video layering method described in Section 3.2 could be applied here to separate the two sources.

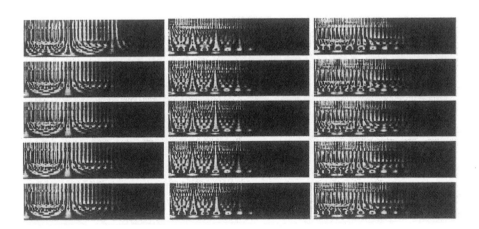

Figure 19.3. Correlogram Frames of Two Audio Sources and the Mixture. Left: of vowel "ah"; Center: of speech "a huge"; Right: of the mixture "ah + a huge". The task is to infer the left and central panel given only the right panel.

Assuming the noise term in Eqn. (2) is zero-mean Gaussian, the prior distributions over the sprite appearances are also Gaussian, and the transformation, appearance map of one signal are independent of those of the second signal, given the correlogram video sequence, probabilistic inference and learning are used to computer a single set of model parameters that represent the entire video sequence. The parameters include the means and variances of the sprite appearance maps and the observation noise variable.

We use the generalized EM algorithm (Neal and Hinton, 1998) to learn the parameters of the model in Eqn. (2). We will then use the learned parameters to infer the correlogram frames of the individual sound source.

5   EXPERIMENTS AND RESULTS

We report our results on separating a mixture of the following two sounds. The first is an utterance of two words "A huge" by a female speaker and the second is an utterance of the vowel "Ah" by a male speaker. The two utterances are both of 5000 samples long with a sampling rate of 8000 Hz. Their wave forms are shown in the top row of Figure 19.5. The single sound mixture is generated by adding these two utterances after they are scaled to have equal energy.

Cochleagram of the sound mixture is generated using an implementation of Lyon's cochlea model by Slaney *et al.* (1994). To produce the 3-D correlogram, each of the 64 cochlea channels undergoes a short-time autocorrelation calculation within a 1000 sample window hopping every 250 samples. Fast Fourier Transform (FFT) is used to calculate this autocorrelation to speed up the calculation. After calculating the correlogram of the sound mixture, we feed each frame of it to the Generative Probabilistic Model to learn the mask, mean and variance of a 2-layer video model.

In this example, since the "Ah" sound is more or less harmonic, its correlogram frames are stable, hence we learn it as a background layer, just like the background images in the video frames shown in Figure 19.2. Its mask is assumed to be uniform with zero variance. However, for the "A Huge" sound, since the fundamental frequency of the speaker changes over time, we need to account for both its change in position and amplitude. We learn it as a foreground layer with unknown mask, like the moving person in the video frames shown in Figure 19.2. As mentioned earlier, the horizontal coordinates of the

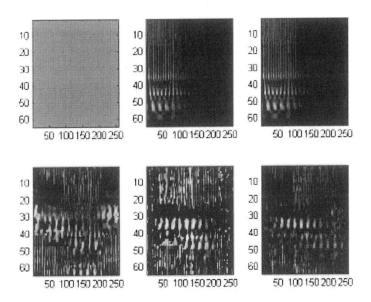

Figure 19.4. Top row: the learned mask, mean of the appearance map and the masked mean of appearance map for the first layer (for the "Ah" sound). Note the third picture is the element-wise product of the first two pictures. Bottom row: those for the second layer (for the "A Huge" speech).

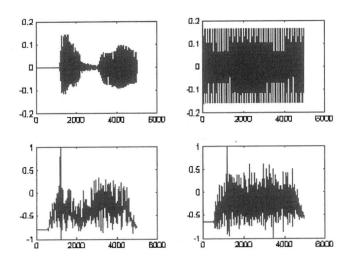

Figure 19.5. Top row: the two original audio sounds, the "A Huge" speech and the "Ah" sound. Bottom row: the separated two sound from a SINGLE mixture of the two sounds in the top row. Although there is large distortion in the separated signals, listening tests show clear separation of the "Ah" sound and the speech sound.

correlogram frames are calculated in the log scale. The learned parameters for the two sounds are plotted in Figure 19.4. Note that the third picture on the top row shows the harmonic structure mostly of the "Ah" sound and the third picture at the bottom row shows the harmonic structure of the "A Huge" sound (e.g., the white stripes in the middle of the image).

The learned mask, mean of each layer and the shift of the mean in each correlogram frame are used to derive two separate correlograms (refer to the video frames for one person and those for another person Figure 19.2). These two correlograms are then used to invert the correlogram generation process to produce two audio-domain signals. The inversion techniques here follow Slaney *et al.* (1994). We shown the wave form of the two re-constructed signals in the bottom row of Figure 19.5. In comparison with the original wave forms shown at the top row, we notice that there exists large amount of distortion. However, our listening tests show clear separation of the "Ah" sound and the "The huge" speech.

There are several issues that contribute to the distortion in reconstruction. First, the phase information for either sound signal is not available, what we have used is the phase of the mixture sound. Next, the correlogram frames of the each sound do not contain localized, rigid objects like the two persons in Figure 19.2. We will address these issues in our future research. Based on the fact the two separated sounds do contain two different audio sources, we believe our proposed approach deserves more research that will lead to promising results.

6   CONCLUSIONS

We have argued the necessity of joint audio-visual scene analysis to deal with the difficult problem of CASA. It is argued that the problem of CASA will benefit from computer audio-visual scene analysis (CAVSA). We also propose a generative probabilistic model on correlogram, the video representation of audio signal, to separate the audio sources.

References

Cooke, M.P., 1993, *Modeling Auditory Processing and Organization*, Cambridge University Press, Cambridge, U.K.

Dempster, A., Laird, N., and Rubin, D., 1997, Maximum likelihood from incomplete data via the EM algorithm, *J. Roy. Statist. Soc. B*, **vol. 39, no. 1**, pp. 1-38.

Frey, B.J. and Jojic, N., 1999, Learning mixture models of images and inferring spatial transformations using the EM algorithm, *Proceedings of the IEEE Computer Vision and Pattern Recognition*, pp. 416-422.

Frey, B.J., and Jojic, N., 2003, Transformation-invariant clustering using the em algorithm, *IEEE Transactions on Pattern Analysis and Machine Intelligence*, **vol. 25, no. 1**, pp. 1-17.

Jojic, N. and Frey, B.J., 2001, Learning flexible sprites in video layers, *Proceedings of the IEEE Computer Vision and Pattern Recognition*, December 2001.

Jojic, N., Petrovic, N., and Huang, T.S., 2003, Scene generative models for adaptive video fast forward, *Proceedings. International Conference on Image Processing*, **vol. 2**, pp. 619-622.

Jojic, N., Petrovic, N., Frey, B.J., and Huang, T.S., 2000, Transformed hidden Markov models: estimating mixture models of images and inferring spatial transformations in video sequences, *Proceedings of IEEE Conference on Computer Vision and Pattern Recognition*, **vol. 2**, pp. 26-33.

Neal, R.M. and Hinton, G.E., 1998, A view of the EM algorithm that justifies incremental, sparse, and other variants, in: *Learning in Graphical Models*, M.I. Jordan, ed., Kluwer Academic Publishers, pp. 355-368.

Pearl, J., 1988, *Probabilistic Reasoning in Intelligent Systems*, Kaufmann, 2nd edition.

Slaney, M., Narr, D., and Lyon, R.F., 1994, Auditory model inversion for sound separation, *Proceedings of the ICASSP 94*, **vol. II**, pp. 77-80.

Tao, H., Kumar, R., and Sawhney, H.S., 2000, Dynamic layer representation with applications to tracking, *Proceedings of the IEEE Conference on Computer Vision and Pattern Recognition*, **vol. 2**, pp. 134-141.

Williams, C., and Titsias, M.K., 2002, Learning about multiple objects in images: Factorial learning without factorial search, *Advances in Neural Information Processing Systems(NIPS)*.

Wilson, A.D. and Bobick, A.F., 1999, Parametric hidden Markov models for gesture recognition, *IEEE Transactions on Pattern Analysis and Machine Intelligence*, **vol. 21, no. 9**, pp. 884-900.

Xiong, Z., Radhakrishnan, R., Divakaran, A., and Huang, T.S., (submitted), Audio-visual sports highlights extraction using coupled hidden markov models, *Pattern Analysis and Application Journal*, Special Issue on Video Based Event Detection.

Chapter 20

Evaluating Speech Separation Systems

Daniel P.W. Ellis
LabROSA, Columbia University
New York NY, U.S.A.
dpwe@ee.columbia.edu

1 THE ASR EXPERIENCE

Quantitative evaluation is an essential and sensitive factor in any area of technological research. Automatic Speech Recognition (ASR) provides an instructive example of the benefits and costs of common evaluation standard. Prior to the mid-1980s, speech recognition research was a confusing and disorganized field, with individual research groups tending to use idiosyncratic measures that showed their particular systems in the best light. Widespread frustration at the difficulty of comparing the achievements of different groups – among researchers and funders alike – was answered by a series of carefully-designed evaluation tasks created by the US National Institute of Standards and Technology (Pallet, 1985). While the speech material in these tasks has evolved from highly constrained vocabularies, grammars and speaking styles through to unconstrained telephone conversations, the principal figure of merit has remained the Word Error Rate (WER) – the number of incorrect word tokens generated by the system as a percentage of the word count of the ideal transcript – throughout this period.

Over more than 15 years of NIST evaluations, the benefits of the common evaluation task and single performance measure have been dramatic. Standard measures have made it possible to give definitive answers to questions over the relative benefits of different techniques, even when those differences are small. Since the advent of Gaussian Mixture Model-Hidden Markov Model (GMM-HMM) recognition systems, it turns out that most ASR improvements have been incremental, rarely affording an improvement of more than 10% relative, yet we have seen a compound system improvement of perhaps two orders of magnitude through the careful and judicious combination of many small enhancements. Accurate and consistent performance measures are crucial to making this possible.

The disadvantage to this powerful organization of the field around a common goal and metric comes from the kind of 'monoculture' we see in current speech recognition research. Of the many hundreds of papers published in speech recognition journals and conference proceedings each year, the vast majority use same GMM-HMM framework or very close relatives, and of the dozen or so labs working on large-vocabulary speech recognition systems and participating in current NIST evaluations, all are using systems that appear identical to a casual observer. If GMM-HMM systems were obviously the 'right' solution, this might be expected; however, many researchers are uncomfortable with the HMM framework, but feel obliged to keep working with it because the performance loss incurred by switching to a less mature, less optimized novel approach would jeopardize the acceptance (and publication) of their work (Bourlard *et al.*, 1996).

The dominance of a common standard can have other disadvantages. The universal adoption of WER as the principal performance measure has led to a focus on transcription tasks and speech-only material to the neglect of other kinds of signals (including, of particular relevance to the current volume, many kinds of speech-interference mixtures). A single style of task and a single performance measure dominating the field for several decades has resulted in solutions more and more closely optimized for that one task, and a widening gap between performance on the focus tasks and other applications, for instance speech mixtures, that may be equally important in a broad sense but happen not to have been included in the evaluations.

1.1   Lessons of evaluation

From the example of speech recognition, we can draw the following lessons:

- Common evaluation tasks (along with the corresponding performance metrics) can have a very positive effect on research and progress in a given field by providing detailed, quantitative answers to questions over the relative merits of different approaches. In addition to furthering debate, this information makes it easier for funding sources to support the field, since they can be more confident that their money is getting results.

- When a single task is defined, and particularly when it bears on funding and other resource allocation, there will be a great concentration on that task leading to the neglect of similar but distinct problems. Thus, the task chosen should ideally represent a real problem with useful applications – so that even in the worst case, with only that one problem being solved, there is still some valuable output from the research.

- Funneling all the effort of a research community into a single, narrow focus is generally undesirable; one alternative is to define more than one task and/or more than one performance measure, to create multiple 'niches' supporting several different threads of research. Making these niches too numerous, however, defeats the benefits of common evaluation: if each separate group evaluates their approach with a different measure, benefits of a common standard are largely lost.

2   EVALUATING SPEECH SEPARATION

An evaluation task consists of two components: a *domain* or application area, specifying the kinds of material that will be considered (such as in-car speech, or target-versus-interferer speech); a *metric* such as word error rate or signal-to-noise ratio, which implicitly defines the core nature of the problem to be addressed (recognizing spoken words or reducing distortion energy respectively). Since certain metrics place additional constraints on the domain (such as the availability of isolated pre-mixture sources), we will first consider the range of metrics that are available and that have been used in speech separation work.

Metrics can be arranged on an axis of abstraction, from those that measure the most concrete, literal properties of signals, through to those concerned with much higher-level, derived properties in the information extracted from the signals.

2.1   Signal-to-noise ratio

The simplest measure, signal-to-noise ratio (SNR), requires that the system being measured reconstructs actual waveforms corresponding to individual sources in a mixture, and that the pre-mixture waveforms of those sources (the 'ideal' outputs) are available. SNR is defined as ratio of the energy of the original target source to the energy of the difference between original and reconstruction – that is, the energy of a signal which, when linearly added to the original, would give the reconstruction. This measure is commonly used for low-level algorithms that have a good chance at near-perfect separation (such as multi-channel Independent Component Analysis (Bell and Sejnowski, 1995), or time-frequency masked reconstruction (Brown and Cooke, 1994)), and is arguably *sufficient*: if we are able to reconstruct a signal that (almost) exactly matches some clean, pre-mixture version, then any other information we wish to obtain is likely also to be available. However, the problems of SNR are:

- It requires the original signal for comparison, largely limiting its use to mixtures that are artificially constructed, rather than those recorded from real environments.

- Distortions such as fixed phase/time delays or nonuniform gains across frequency which can have only a small effect on the perceived quality of a reconstructed sound, can have a large negative effect on SNR.

- The common unit of measurement, energy, has in general only an indirect relationship to perceived quality. The same amount of energy will have a widely-varying impact on perceived quality depending on where and how it is placed in time-frequency; this is particularly significant in the case of speech, where *most* of the energy is below 500 Hz, yet very little intelligibility is lost when this energy is filtered out. Another example of the disconnect between SNR and perceived quality comes from the psychoacoustic-based coding used in schemes like 'MP3' audio, where a reproduction with an SNR of under 20 dB can sound essentially perfect because all the distortion energy has been carefully hidden below the complex masking patterns of the auditory system.

2.2   Representation-based metrics

While SNR has the attraction of being applicable to any system that generates an output waveform, more helpful measures (at least from the point of view of system development) can be derived directly from whatever representation is used within a particular system. Thus, in Cooke's original Computational Auditory Scene Analysis (CASA) system (Cooke, 1991), an evaluation was performed by comparing the 'strands' representations resolved by his system with the representation generated for each source in isolation, thereby avoiding the need for a strands-to-sound resynthesis path.

By considering the internal representation, evaluations can also be made relative to an 'ideal' performance that reflects intrinsic limitations of a given approach. Many recent systems are based on time-frequency (TF) masked refiltering, in which Gabor 'tiles' in TF are classified as target-dominated and selectively resynthesized (Hu and Wang, 2003; Roweis, 2001). Such an approach cannot separate overlapped energy falling into a single cell, so an SNR ceiling is achieved by an 'ideal' mask consisting of all the cells in which target energy is greater than interference (since including any other cells will increase the distortion energy, by adding noise, more than it is decreased by reducing the deleted target). Systems based on these masks can be evaluated by how closely they approach this ideal mask e.g. by measuring the classification accuracy of TF cells. This measure removes the effective weighting of each cell by its local signal energy in SNR calculation; however, it gives a disproportionate influence to near-silent TF cells whose 'ideal' classification will depend on the likely irrelevant noise-floor level in the original mixture components.

Another analysis possible with masked refiltering systems is the separate accounting for distortion due to energy deleted from the target, and due to included portions of the interference (the "energy loss" and "noise residue" of (Hu and Wang, 2003)). However, we are again faced by the problem of the perceptual incomparability of energy in different parts of time-frequency.

2.3 Perceptual Models

As indicated above, the inadequacies of SNR have long been apparent in the active and successful field of audio coding. When the goal is to satisfy human listeners (e.g. telephony customers or music consumers), there is no substitute for formal listening tests in which subjects rate the perceived quality of various algorithms applied to the same material. Due to the cost of such evaluations, however, considerable effort has gone into developing algorithmic estimates such as the ITU standards for PEAQ and PESQ (Perceptual Evaluation of Audio/Speech Quality (Thiede *et al.*, 2000)). While these measures also require a pre-distortion original reference, they make sophisticated efforts to factor out perceptually-irrelevant modifications, and their use for the evaluation of low-level signal-separation systems deserves investigation.

2.4 High-level attributes

While perfect signal recovery may be sufficient, it is rarely necessary. Signal separation is not an end in itself, but a means to some subsequent application, be that recognizing the words in some noisy speech, or even the pleasure of listening to a solo musical performance without additional instruments. In every case, metrics can be devised to measure more directly the success of the signal separation stage on the overall application. When the ultimate task is extracting specific parameters, such as the times of occurrence of certain events, or perhaps limited descriptions of such events (such as onset times, pitches, and intensities in polyphonic music transcription), it is natural to evaluate in terms of the error in that domain.

By far the most widespread evaluation falling into this category is word error rate of speech recognition systems for mixed signals. Given the widespread acceptance of WER as a measure for isolated speech recognition, it is natural to extend the same metric to conditions of significant interference, even when substantially different processing is introduced to address that interference. This approach was taken in one of the earliest models of auditory scene analysis (Weintraub, 1985), although in that work, as in many subsequent experiments, it was found that in some cases that the signal separation preprocessing made the error rate worse than simply feeding the original mixture to an unmodified recognition engine.

Using signal separation to aid speech recognition requires a careful match between the separation techniques and the recognition engine: one example of a well-matched combination highlights the difference between this and lower-level metrics. In missing-data recognition (Cooke *et al.*, 2001), the matching between observed signal and learned speech models is modified to account for limited observability of the target i.e. that certain dimensions can be missing at different times. Acoustic scene organization algorithms are then employed to indicate which dimensions (e.g. TF cells) are reliable correlates of the target speech at each instant (Barker *et al.*, 2004). This work reports minimal increase in word error rate in cases when significantly less than half the features are deemed 'reliable' – a situation which would likely give a highly-distorted resynthesis, but which still contains plenty of information to recognize the spoken words largely without ambiguity.

However, because of the sensitivity of WER measures to the compatibility (and, as in (Barker *et al.*, 2004), close functional integration) between separation algorithm and speech recognizer, this measure is only appropriate for systems specifically built for this application.

2.5   Domains

Among acoustic signal separation tasks, speech mixed with different kinds of interference is the most popular domain and is our main concern here. The target voice can experience different amounts of spectral coloration, reverberant smearing, or other distortion, but the main axis of variation is in the nature of the interference signal. Simplest is Gaussian white noise, which can be made a more relevant masker by filtering into pink noise (equally energy per octave) or to match some average speech spectrum. In speech recognition, a common approach is to train models for the combination of speech-plus-noise, which can be very successful for such stationary noise particularly when the absolute level is constrained; a distinct stage of signal separation is avoided completely.

Real-world sound sources comprise a more challenging form of interference because their impact on the features cannot be predicted so accurately (i.e. with small variance), although the combination of a large number of independent sources will tend towards Gaussian noise. At a given power level, the most difficult interference should be a single second voice, since the statistical features of the interference are indistinguishable from the target. (In practice, a single voice offers many opportunities 'glimpsing' the target during silent gaps in the interference, so a combination of a small number of unsynchronized voices may achieve greater interference.)

The majority of noisy speech tasks are created artificially by mixing speech recorded in quiet conditions with different "pure interference" signals (e.g. (Pearce and Hirsch, 2000)). This approach has the attractions that the relative levels of speech and noise can be adjusted, the same speech can be embedded in several different types of noise, and the clean speech can be used to generate a baseline performance. However, it is a poor match to reality: there is no guarantee that the synthetic mixture actually resembles something that could ever have been recorded in a noisy environment, not least because of the Lombard effect, the reflexive modification of speaking quality employed by humans to overcome noisy environments (Lane and Tranel, 1971). Other effects such as reverberation are also frequently ignored in synthetic noise mixtures.

Given the problem identified above of solving only what we test, it would seem preferable to use real recordings of speech in noisy environments as test material. While some data of this kind do exist (Schmidt-Nielsen *et al.*, 2000), on the whole it is avoided due to the flexibility and control available with synthetic mixtures as just mentioned: recording a range of speech material against a range of background noise types and levels requires, in the worst case, a separate recording for each condition, rather than factorized combinations of a few base recordings.

Another problem with real-world recordings is the availability of ground-truth descriptions. If we artificially mix a clean voice signal against a noisy background, we may hope that our speech separation algorithms will recreate the original clean speech; if the noisy speech is all we have, how can we even judge the quality of the resynthesis? I would argue, however, that this assumption that the pre-mixture original represents the unique best output we could hope for is in fact dodging the more difficult, but more important, question of deciding what we *really* want. If the purpose of the algorithm is to enhance noisy speech for a person to listen to, then the appropriate metric is subjective quality rating, not similarity to an original signal which may not, in fact, match the impression of the source speech in the mind of the listener.

This appeal to subjective sources for ground truth in complex mixtures extends beyond speech in noise: In (Ellis, 1996), a computational auditory scene analysis system that sought to mark the occurrence of different sound events in complex, real-world ambient sounds was evaluated on its ability to duplicate the consensus results of a set of listeners given the same task.

3   CONCLUSIONS AND RECOMMENDATIONS

In light of this discussion, we make some recommendations for the form of a future, widely-applicable evaluation task for speech separation systems:

- It should be based on some kind of real-world task, so that if the worst-case occurs and we end up with solutions applicable only to this narrow task, they can at least be deployed to some purpose.

- The data should be real recordings, or possibly synthetic recordings in which all the possibly relevant aspects of the real recording have been carefully duplicated.

- The evaluation ground truth (be it word transcripts, event detection and descriptions, or other information from the signal) should originate from human transcribers to get at the 'subjective' character of the sound.

- As this implies, the domain of comparison should be in terms of high-level information and attributes, rather than low-level comparisons against some ideal waveform.

- If the task represents a real and useful domain, it ought to be possible to gather comparable human performance on the same task, so we can accurately measure how well our machines do relative to the best currently-known listening machine. Ideally, this would be a task that humans (perhaps impaired populations) find somewhat difficult, to give the machines a chance to exceed human performance – although machines that came anywhere close to human performance on any kind of acoustic scene analysis would be welcome.

One possible domain is audio recorded in real, multi-party meetings, and this task has recently begun to attract attention (Yu *et al.*, 1999; Morgan *et al.*, 2001). Such corpora typically involve significant amounts of speech overlap, and often have both near- and far-field microphone recordings; the head-mounted near-field mics provide a kind of ground-truth reference for the voices picked up by the far-field tabletop mics.

Speech separation is often referred to as the Cocktail-Party problem (following (Cherry, 1953)), and a room containing multiple simultaneous conversations might provide an interesting test domain, one that would mostly defeat human listeners. Such a party could be staged with each participant wearing a head-mounted microphone (which can be inconspicuous) to provide some level of ground-truth. An interesting corpus along these lines is the Sheffeld-ATR Crossword task (Crawford *et al.*, 1994), which involved two simultaneous conversations with a fifth participant occasionally involved in both.

A final area is the kind of continuous personal recording proposed in (Bush, 1945) and investigated in (Clarkson *et al.*, 1998): wearable microphones and miniature hard-disk recorders can easily make complete records of a user's acoustic environment, but to allow any kind of useful retrieval from hundreds of hours of such recordings requires automatic analysis of which acoustic source separation will be an important part.

In conclusion, insights from speech recognition and elsewhere show that a common evaluation task is critical to the future progress and support of speech separation research. The form and nature of such a task, however, is far from clear, not least because there is little consensus on the real purpose or ultimate application for speech separation technologies. We favor a task firmly embedded in real-world scenario, and an evaluation metric that reflects subjective information extraction rather than an objective, but arbitrary, low-level ideal.

References

Barker, J., Cooke, M., and Ellis, D., 2004, Decoding speech in the presence of other sources, *submitted to Speech Communication*.

Bell, A.J. and Sejnowski, T.J., 1995, An information- maximization approach to blind separation and blind deconvolution, *Neural Computation*, **7(6)**:1129–1159.

Bourlard, H., Hermansky, H., and Morgan, N., 1996, Towards increasing speech recognition error rates, *Speech Communication*, pages 205–231.

Brown, G. J. and Cooke, M., 1994, Computational auditory scene analysis, *Computer speech and language*, **8**:297–336.

Bush, V., 1945, As we may think. *The Atlantic Monthly*.

Cherry, E.C., 1953, Some experiments on the recognition of speech with one and two ears. *J. Acoust. Soc. Am.*, **25**:975–979.

Clarkson, B., Sawhney, N., and Pentland, A., 1998, Auditory context awareness via wearable computing, in *Proc. Perceptual User Interfaces Workshop*.

Cooke, M.P., 1991, *Modelling auditory processing and organisation*. Ph.D. thesis, Department of Computer Science, University of Sheffeld.

Cooke, M., Green, P., Josifovski, L., and Vizinho, A., 2001, Robust automatic speech recognition with missing and unreliable acoustic data, *Speech Communication*, **34(3)**:267–285.

Crawford, M.D., Brown, G.J., Cooke, M.P., and Green, P.D., 1994, Design, collection and analysis of a multi-simultaneous-speaker corpus, In *Proc. Inst. Acoustics*, **volume 5**, pages 183–190.

Ellis, D.P.W., 1996, *Prediction–driven computational auditory scene analysis*. Ph.D. thesis, Department of Electrical Engineering and Computer Science, M.I.T.

Hu, G. and Wang, D.L., 2003, Monaural speech separation, in *Advances in NIPS 13*, Cambridge MA. MIT Press.

Lane, H. and Tranel, B., 1971, The Lombard sign and the role of hearing in speech. *J. Speech and Hearing Res.*, (**14**):677–709.

Morgan, N., Baron, D., Edwards, J., Ellis, D., Gelbart, D., Janin, A., Pfau, T., Shriberg, E., and Stolcke, A., 2001, The meeting project at ICSI, in *Proc. Human Lang. Tech. Conf.*, pages 246–252.

Pallet, D.S., 1985, Performance assessment of automatic speech recognizers. *J. Res. Natl. Bureau of Standards*, **90**:371–387.

Pearce, D. and Hirsch, H.-G., 2000, The AURORA experimental framework for the performance evaluation of speech recognition systems under noisy conditions, in *Proc. ICSLP '00*, **volume 4**, pages 29–32, Beijing, China.

Roweis, S., 2001, One-microphone source separation, in *Advances in NIPS 11*, pages 609–616. MIT Press, Cambridge MA.

Schmidt-Nielsen, A., Marsh, E., Tardelli, J., Gatewood, P., Kreamer, E., Tremain, T., Cieri, C., and Wright, J., 2000, Speech in Noisy Environments (SPINE), *Evaluation Audio. Linguistic Data Consortium*, Philadelphia PA.

Thiede, T., Treurniet,W.C., Bitto, R., Schmidmer, C., Sporer, T., Beerends, J.G., Colomes, C., Keyhl, M., Stoll, G., Brandeburg, K., and Feiten, B., 2000, PEAQ – the ITU standard for objective measurement of perceived audio quality, *J. Audio Eng. Soc.*, **48**(1/2).

Weintraub, M., 1985, *A theory and computational model of auditory monoaural sound separation*. Ph.D. thesis, Department of Electrical Engineering, Stanford University.

Yu, H., Finke, M., and Waibel, A., 1999, Progress in automatic meeting transcription.

Chapter 21

Making Sense of Everyday Speech: a Glimpsing Account

Martin Cooke
Department of Computer Science
University of Sheffield
m.cooke@dcs.shef.ac.uk

1 INTRODUCTION

To make sense of speech in everyday conditions, listeners have to cope with distortions produced by additive noise, reverberation and channel characteristics. A better understanding of the processes involved will lead to progress in robust automatic speech recognition and will inform the design of more selective hearing aids. After several decades of work on both the psychophysics and algorithmics of speech separation, a clearer picture of the factors involved in the process of understanding speech in the presence of other sources is emerging.

The first part of this paper describes the main factors which contribute to our ability to understand speech in noise. Some of the many interactions between such factors are highlighted and the implications for models of speech segregation are discussed. The second part of the paper introduces a model for the perception of noisy speech based on the notion that listeners exploit 'glimpses' of the target signal rather than attempting to completely separate out a clean speech signal. A study which explores the issues of sufficiency and utility of glimpses is discussed.

2 SOURCE UNDERSTANDING IN NOISE

Many factors can influence a listener's ability to make sense of a target source such as speech in an acoustic mixture. Some of these are listed below:

1. Audibility of the target (energetic masking)
2. Confusability of the target and background (informational masking)

3. Availability of organisational cues in the target

4. Availability of organisational cues in the background

5. Existence of schemas for the target

6. Existence of schemas for the background

The list is not complete, since factors such as attention and cues from other modalities (principally vision) also play a role in source understanding.

These factors do not act independently. The following list describes some of the many interactions between them.

- Background cues may help define the target via cancellation, while target cues can define the background as a residue (3, 4). Both cancellation and residue-based strategies have been proposed in models of auditory scene analysis (de Cheveigné, 1993; Nakatani *et al*, 1998).

- Target and background cues may act separately or jointly to detect coherent audible regions arising from one or other source in a mixture (1, 3, 4). Bottom-up approaches to CASA (e.g. Brown and Cooke, 1994) employ precisely these types of cues in source segregation.

- While the preceding suggests that background grouping cues may be useful, an absence of grouping cues in the background sources may reduce the amount of informational masking (2, 6).

- Differences in target and background grouping cues may appear as a release from energetic masking (1, 3, 4). For example, onset differences, which are known to help in source segregation (Darwin, 1981; Bregman, 1990), also cause a reduction in spectral overlap and hence improve the audibility of source components.

- Differences in target and background properties may decrease informational masking (2, 3, 4). For example, a difference in F0 contour may make sources less confusable.

- Target and background schemas may co-operate to jointly explain the observed mixture (5, 6). This approach underlies so-called 'model-based' techniques for source segregation such as HMM decomposition (Varga and Moore, 1990) and MAXVQ (Roweis, 2003).

- Schemas for the target may be used to detect and integrate audible regions, with or without the help of other factors (1, 5). In a sense, this is the approach championed by Remez *et al* (1994), who argue that speech is sufficiently special to allow learned representations of the signal to be employed in extracting speech from mixtures.

- Poor discriminability between target and background schemas might appear as informational masking (2, 5, 6). For example, a lack of exposure to certain sources (perhaps while learning a second language) is likely to make them more confusable.

- Factors which appear as bottom-up, organisational cues may in fact be embedded in schemas via learning (3, 5). For example, cues which arise from vocal tract length differences or different pitches may be represented as sets of schemas for each speech unit.

The interactions detailed above have a number of consequences for research in sound segregation. First, there is clearly the potential for multiple explanatory mechanisms, as evidenced by the variety of algorithms which have been proposed. Some of these can be regarded as wholly 'bottom-up' and others as wholly 'top down', but this distinction is oversimplified given the space of possible interactions. Second, certain processes may masquerade as others. For instance, cue differences may lead not only to a reduction in confusability but at times to a reduction in energetic masking. Third, tradeoffs exist: for example, harmonic backgrounds increase the opportunities for cancellation strategies, but may also lead to an increase in informational masking of harmonic foregrounds.

This analysis raises a number of issues:

- What are the relative contributions of these factors in sound source separation and understanding?

- What is the role of background sources? Does it help to have schemas for these sources? Does it help if the background contains grouping cues?

- Can successful identification occur without foreground and background grouping cues?

- How fine-grained is separation? Do listeners separate in order to identify, or is separation a by-product, or even an illusion?

Current and recent work is clarifying some of these issues, at least for speech targets. For instance, Brungart and colleagues are exploring the relative roles of informational and energetic masking in cocktail-party-like situations (Brungart *et al*, 2001).

3   AN ILLUSTRATION: SINGLE COMPETING TALKER VS MULTI-SPEAKER BABBLE

The interplay of the factors introduced in section 2 can be illustrated by considering their explanatory power in the case of speech masked by N-speaker babble, for N=1 and 8. Figure 21.1, from Miller (1947), demonstrates that when the source and background are equally intense, listeners have a clear preference for a single competing talker, in spite of the fact that the degree of informational masking in this situation is much stronger than in the multispeaker babbler case. In fact, one might expect the multispeaker babble to present fewer problems for source separation for another reason: it is a less variable source. It is well-known that computational approaches to speech enhancement provide little or no benefit in the face of even moderate departures from background stationarity. Algorithmically, removal of quasi-stationary noise presents few problems, but it appears that listeners employ other strategies.

In terms of the factors outlined in section 2, there are at least three ways in which listeners might benefit from a single-speaker background. First, they are able to employ organisational cues for both foreground *and* background sources, leading to the possibility for cancellation-based enhancement, at least for the voiced parts. Second, they have access to models for both the foreground and background sources, allowing the operation of processes which jointly attempt to explain the observed energy at each time-frequency point. Third, the variability in the speech masker allows for a greater number of opportunities to 'glimpse' the target when compared to the multispeaker babble case. This final factor is illustrated in Figure 21.2, which shows the percentage of time-frequency 'pixels' which have a locally-favourable SNR for the speech target for both single-speaker and 8-speaker backgrounds.

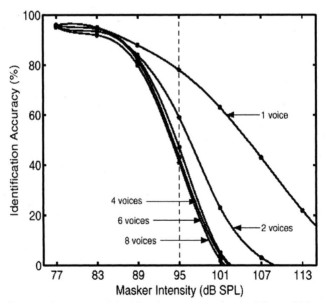

Figure 21.1. Listeners' consonant identification accuracy in conditions of N-speaker babble. Redrawn by Assmann and Summerfield (in press) from Miller (1947).

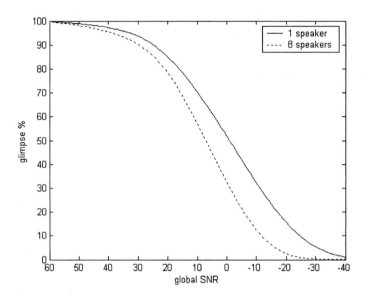

Figure 21.2. Proportion of glimpses of a speech target available in the presence of a background composed of 1 or 8 speakers, as a function of SNR.

The relative contribution of the three factors to the single-speaker background advantage is at present unclear. While perceptual models inspired by the first two factors – grouping and schemas – have appeared, less attention has been paid to the role of glimpsing. The following section outlines a model inspired by the glimpsing notion.

4   A GLIMPSING MODEL

4.1   Why glimpse?

The notion that listeners can utilise glimpses of the target source in everyday noise backgrounds contrasts with the source enhancement approach, in which listeners attempt to recover the target source prior to identification. The latter approach has dominated the computational literature on robust automatic speech recognition, primarily since it allows the use of unmodified recognition technology. However, Cooke *et al* (1994; 2001) and others have demonstrated that modifying the recognition process to admit the possibility of 'missing data' leads to substantial improvements in robust ASR performance. Glimpsing can be thought of as the perceptual equivalent of missing data techniques in ASR. Cooke (2003a) contains detailed arguments for glimpsing in speech perception.

A working definition of a glimpse is *some time-frequency region which contains a reasonably undistorted 'view' of local signal properties*. For concreteness, we can imagine that the local signal property of interest is energy, although it is likely that other properties such as local estimates of F0, AM and FM will be useful too. Is it reasonable to expect such undistorted views of speech targets in natural conditions? Fortunately, the answer is yes, due to two factors: the log-like compression of energies in the peripheral auditory system, and the modulation frequencies present in speech due to the patterning of voiced, unvoiced and silent segments. The former property ensures that the log energy at virtually all spectro-temporal points is dominated by the stronger source, while the latter provides the nonstationarity which increases both the number and extent of glimpses. Figure 21.3 illustrates both properties in action for a speech target masked by a single-speaker background. The partitioning of a mixture on the basis of energetic dominance is almost perfect, with very few 'ambiguous' regions. Further, glimpses of one or other source tend to be quite large and coherent.

4.2   Issues for a glimpsing model

While glimpsing based on mixture partitioning appears to be an attractive alternative to complete speech separation, it presupposes the solution to a number of problems:

- Is *sufficient information* contained in glimpses to support identification?
- What constitutes a *useful glimpse*?
- How might listeners and computational techniques *detect* and *integrate* glimpses?

The study described in the following section addressed the first two questions. Glimpse detection remains an open problem, though techniques based on computational auditory scene analysis (Cooke and Ellis, 2001) are likely to be part of the solution. Glimpse integration is being tackled via a new form of speech decoder described in Barker *et al* (in press).

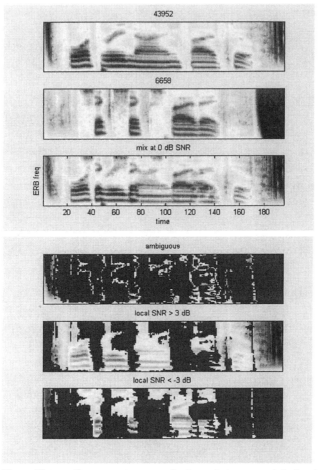

Figure 21.3. Upper Plot: auditory spectrograms for two utterances and their mixture at 0 dB SNR (bottom). Lower Plot: the middle and lower panels show those regions in the mixture which are dominated by one or other source, where dominance is defined as having a local SNR exceeding 3dB. The upper panel depicts 'ambiguous' regions whose local SNR is in the range (-3, 3) dB.

4.3 Glimpsing in VCVs: listeners vs a model

To measure the extent to which listeners can utilise glimpses in noisy speech perception, subjects and a computational model based on missing data techniques were compared on a consonant identification task (Cooke 2003b). Subjects were asked to distinguish 16 consonants presented in an /a/_/a/ context, using the corpus collected by Shannon *et al* (1999). Two noise backgrounds consisting of reversed N-speaker babble, for N=1 and 8 were employed, at three target-to-masker ratios (0, -6, -12 dB). Figure 21.4 (left) shows listeners' identification rates. These results bear out the 8-12 dB advantage of single-speaker backgrounds over multi-speaker babble found by Miller (1947). Figure 21.4 (left) also shows the performance of an 'ideal' glimpsing model which utilises *all* spectro-temporal regions with a locally-favourable SNR. Clearly, there is more than enough information in the glimpsed regions to explain listeners' performance on this task.

How likely is it that listeners can detect all possible glimpses, given that some will occupy a very small region of the spectrogram and will be swamped in a sea of noise? A 'useful' glimpse can be defined as one which has at least a certain extent in time and frequency. By searching through all possible minimum (rectangular in time-frequency) extents, a good match to listeners' performance could be obtained for a minimum size of 6.3 ERBs by 40 ms. Figure 21.4 (right) shows the model performance in this case.

5   DISCUSSION

One interpretation of these results is that glimpsing alone provides a potential explanation for listeners' performance on this task, but these results should be interpreted with caution. First, the masking model which underlies the choice of glimpses does not take nonsimultaneous masking into account. Second, unpublished experiments have demonstrated that listeners perform almost as well in the presence of speech modulated noise as they do in natural speech backgrounds, in spite of the fact that speech modulated noise provides fewer glimpses of the target. In terms of the factors outlined in section 2, one might interpret this result as the consequence of a release from informational

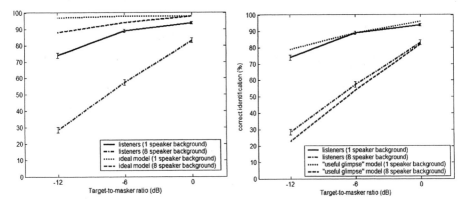

Figure 21.4. Left: listeners versus ideal model performance on a VCV identification task for both single and eight speaker maskers. Right: comparison of listeners with a model which assumes a minimum useful glimpse size of 6.3 ERBs in frequency by 40 ms in time.

masking, although the absence of background grouping cues or schemas would be expected to lead to a reduced performance. Further work is required to determine the relative contributions of these factors.

How do glimpses fit in with the energetic and informational masking concepts? Superficially, glimpsing appears to be closely related to the former. However, it would be wrong to equate the glimpsing account solely with energetic masking. While energetic masking provides a baseline defining those spectro-temporal regions which might constitute glimpses, the ability to make sense of speech from these regions will depend on mechanisms for their detection, and, crucially, their integration. In the 2 speaker case, we can regard glimpse integration as akin to making sense of a jigsaw image where the pieces of two jigsaws are not only jumbled up, but where some significant proportion is lost in the process. The factors which determine whether this integration problem is easy to solve may very well be similar to those thought to be implicated in informational masking, viz. similarity in properties such as F0, vocal tract size and phonemic class.

References

Assmann, P. and Summerfield, Q. (in press) The perception of speech under adverse acoustic conditions. In *Speech Processing in the Auditory System, Springer Handbook of Auditory Research Vol. 14*, S. Greenberg and W. Ainsworth, eds., Springer.

Barker, J., Cooke, M.P., and Ellis, D.P.W. (in press) Decoding speech in the presence of other sources, *Speech Communication.*

Bregman, A.S., 1990, *Auditory Scene Analysis.* Cambridge, MA: MIT Press.

Brown, G.J. and Cooke, M.P., 1994, Computational auditory scene analysis, *Computer Speech and Language*, **8**, 297-336.

Brungart, D.S., Simpson, B.D., Ericson, M.A., and Scott, K.R., 2001, Informational and energetic masking effects in the perception of multiple simultaneous talkers, *J. Acoust. Soc. Am.,* **110**, 2527-2538.

de Cheveigné, A., 1993, Separation of concurrent harmonic sounds: Fundamental frequency estimation and a time-domain cancellation model of auditory processing, *J. Acoust. Soc. Am.*, **93**, 3271-3290.

Cooke, M.P., 2003a, Glimpsing Speech, *Journal of Phonetics.*

Cooke, M.P., 2003b, A glimpsing model of speech perception, *Proc. Int. Cong. Phonetic Sciences.*

Cooke, M.P., Green, P.D., and Crawford, M.D., 1994, Handling missing data in speech recognition. *Proc. ICSLP,* 1555-1558.

Cooke, M.P., Green, P.D., Josifovski, L., and Vizinho, A., 2001, Robust automatic speech recognition with missing and uncertain acoustic data, *Speech Communication*, **34**, 267-285.

Cooke, M.P. and Ellis, D.P.W., 2001, The auditory organization of speech and other sources in listeners and computational models, *Speech Communication*, **35**, 141-177.

Darwin, C.J., 1981, Perceptual grouping of speech components differing in fundamental frequency and onset-time, *Quarterly Journal of Experimental Psychology*, **3A**, 185-207.

Miller, G.A., 1947, The masking of speech, *Psych. Bull.*, **44**, 105-129.

Nakatani, T., Okuno, H.G., Goto, M., and Ito, T., 1998, Multiagent based binaural sound stream segregation, in: *Readings in Computational Auditory Scene Analysis,* D. Rosenthal and H. Okuno, eds., Lawrence Erlbaum.

Remez, R.E., Rubin, P.E., Berns, S.M., Pardo, J.S., and Lang, J.M., 1994, On the perceptual organization of speech, *Psychological Review*, **101**(1), 129-156.

Roweis, S.T., 2003, Factorial models and refiltering for speech separation and denoising, *Proc. Eurospeech*, 1009-1012.

Shannon, R.V., Jensvold, A., Padilla, M., Robert, M.E., and Wang, X., 1999, Consonant recordings for speech testing, *J. Acoust. Soc. Am.*, **106**, L71-L74.

Varga, A.P. and Moore, R.K., 1990, Hidden Markov Model decomposition of speech and noise, *Proc. ICASSP*, 845-848.

Index

A

Acoustic Scene Analysis 241, 252, 298
Aging 274, 276
AIM (Auditory Image Model). See Modeling.
AM (Amplitude Modulation) 89, 178, 187, 266–267, 274–275, 306
Aperiodicity. See Periodicity.
Articulatory Modeling. See Modeling.
ASA. (Auditory Scene Analysis) 3, 5, 7, 9, 11, 13–21, 26, 32, 35–36, 47, 58, 62, 96, 177–179, 181, 183, 191, 195, 200, 202, 205, 217, 241, 249, 275, 294, 296–297, 302, 304
ASR (Automatic Speech Recognition) 31, 33, 72–73, 133–134, 141, 145, 148, 177, 179–180, 182, 185, 190–191, 203, 209–211, 213, 241–242, 246, 251, 254, 261, 291, 301, 305–306
Audio-Visual Scene Analysis 279–280, 288
Auditory Image 43–44, 153–154, 157–158
Auditory Image Model. See Modeling.
Auditory Masking. See Masking.
Auditory Modeling. See Modeling.
Auditory Perception 8, 26, 84, 133, 148, 164, 168, 181–182, 197, 200, 205, 218, 221, 230
Auditory Processing 5, 32, 35, 38, 85, 93, 143, 213, 219, 230

B

Bayesian Classification 55, 59, 189, 280–281
Beamforming 61, 65–68, 71–76, 78–80, 145, 244, 250, 253
Binaural Segregation. See Segregation
Binaural Hearing 143
Blind Source Separation See Separation.

C

Cancellation 66, 215, 241, 243–248, 250–253, 302–304
CASA (Computational Auditory Scene Analysis) 58, 62, 99, 137, 141, 148–149, 178–184, 187, 189, 191, 195, 198, 205, 249, 275, 279, 288, 294, 302